The Science of Living Longer

The Science of Living Longer

Developments in Life Extension Technology

Gini Graham Scott, PhD

An Imprint of ABC-CLIO, LLC

Santa Barbara, California • Denver, Colorado

Library of Congress Cataloging-in-Publication Data

Names: Scott, Gini Graham, author.
Title: The science of living longer : developments in life extension technology / Gini Graham Scott, PhD.
Description: Santa Barbara, California : Praeger, an Imprint of ABC-CLIO, LLC, [2018] | Includes bibliographical references and index.
Identifiers: LCCN 2017030569 (print) | LCCN 2017032276 (ebook) | ISBN 9781440857157 (ebook) | ISBN 9781440857140 (set : alk. paper)
Subjects: LCSH: Longevity. | Life spans (Biology) | Biotechnology.
Classification: LCC QP85 (ebook) | LCC QP85 .S36 2018 (print) | DDC 612.6/8—dc23
LC record available at https://lccn.loc.gov/2017030569

ISBN: 978-1-4408-5714-0 (print)
        978-1-4408-5715-7 (ebook)

22 21 20 19 18   1 2 3 4 5

This book is also available as an eBook.

Praeger
An Imprint of ABC-CLIO, LLC

ABC-CLIO, LLC
130 Cremona Drive, P.O. Box 1911
Santa Barbara, California 93116-1911
www.abc-clio.com

This book is printed on acid-free paper ∞

Manufactured in the United States of America

Contents

ı

Introduction

The Science of Living Longer is about the long search for immortality and extended life throughout history. It describes the scientific and techno-logical developments making this possible, the growing community of people who are part of this search, and how these developments will affect society.

The Long Quest for Immortality

While the search for immortality dates back to prehistoric times, in the past this search was framed more in philosophical, spiritual, and religious terms. It was based on the notion that there is a spiritual essence or soul that is separate from the body, which lives on when the physical body dies. In the Christian tradition, the belief is that the individual who has lived a good, moral life will go to heaven; in the Islamic faith, there is a belief in a kind of paradise; in the Hindi tradition, the spiritual being is reborn again and again in other forms based on the law of karma. So, according to this law, this spiritual essence goes on to live in other forms or bodies, depending on how well a person has lived his or her life, until the soul has achieved the highest form of existence, called nirvana. At this point, there is no more need for rebirth, and the soul becomes a kind of pure essence. Likewise, the Egyptians developed a number of funeral and burial practices, based on preparing the individual for an eternal life beyond the death of the physical body. Many other examples can be found in other cultures and throughout history of this belief in an immortal essence that lives on after death, whether in physical or spiritual form. Indeed, the belief in an afterlife is a fundamental principle in most religions. Besides Christianity, Islam, Hinduism, and Egyptian theology, it can be found in Judaism, Buddhism, Jainism, Sikhism, Zoroastrianism, Judaism, and the Baha'i faith.

The lengths to which humans have gone to achieve immortality has ironically resulted in death for those who tried, like Icarus falling out of the sky when he tried to fly too close to the sun on wings made of feathers and wax. As another example, in ca. 200 B.C.E., the first emperor of China, Qin Shi Huang poisoned himself with a fatal dose of mercury when he ate some mercury pills, thinking these would give him eternal life.[1] More than 1,000 years later, in 1492, Pope Innocent II died when he had a blood transfusion from three healthy boys, in an attempt to absorb their youth. Later, in 1868, Kentucky politician Leonard Jones claimed he had achieved immortality by prayer and fasting when he ran for the U.S. presidency and asserted in his platform that he would give his "secrets for cheating death to the public" upon his election. Even so, that wasn't enough to win, and in a kind of karmic backfire, he died of pneumonia later that year.[2] So much for achieving immortality.

Even before recorded history, humans developed spiritual beliefs in the continued existence of the individual spirit, reflected in the burial practices of many preliterate cultures. Even the Neanderthals seemed to have some kind of belief in a continued spiritual essence, reflected in the discoveries of burial items in what appear to be graves of deceased individuals in their caves.

In virtually all cultures, the existence of ghosts that come back to haunt individuals or remain where they have unfinished business or memories is another way of expressing this belief in immortality; so is the belief in many cultures, such as the Chinese, that ancestors must be honored or placated, lest they return to express their displeasure, resulting in bad fortune. Indeed, the notion of being able to communicate with the spirits of the dead is based on the notion that immortality in some non-physical form exists, and this belief formed the basis of the popular spiritual movement that swept through America in the late 19th century. It also underlies the continued practice of séance to contact the spirit of someone who has died.

This quest for actual immortality for the living has a long history as well. In the 16th century, Ponce de Leon, who became the first governor of Puerto Rico, led an expedition in 1513 to the coast of what became Florida, to find a rumored Fountain of Youth on an island now known as Bimini.[3] And even before and since then, humans have looked for some way to ward off death and retain their continued youth. As Laurie-Ann Vazquez points out in "How Science Is Making Immortality a Reality," the drive behind the quest for the Fountain of Youth—finding the cure for aging—has long been with us. To this end she notes, "humans have been trying to crack the code of immortal youth for almost as long as we've been alive. We've tried just about everything we can imagine, from magic objects and epic journeys, to human sacrifice and drinking blood (and inventing monsters that live forever by doing so)."[4]

The story of Count Dracula, a debonair aristocrat who survives the centuries by drinking the blood of his victims, and the many modern permutations on this story in books and films, certainly play on this desire for immortal life. These books and films include: the *Twilight Series,* originally written by Stephenie Meyers; the *Vampire Diaries,* based on novels by Ann Rice; and the *Walking Dead* and other zombie movies. However, these books and films have a gothic twist, such as the vampire's need to suck the blood for immortal life and the zombie's bite that both kills its victim and turns it into a zombie. An underlying theme in these stories is that the bite of Dracula or a zombie can eventually threaten all humanity, since potentially everyone might become a vampire or a zombie if the menace of these immortal dead beings can't be stopped.

The Possibility of Immortality

Given this long desire for immortality, can we really live forever? Or can we at least experience a greatly extended lifespan? Can aging be stopped or reversed? Increasingly, people are asking these questions as society ages, and scientists and technologists find more and more methods to make a very long life and even immortality possible. So far, scientists and technologists may not have cracked the immortality code, but they are part of a growing quest, which is supported by billions of dollars in mostly private funding by wealthy investors to make immortality and a much longer life span possible.

Of course, accidents, illnesses, injuries, homicides, and the many other ways that people suddenly and unexpectedly die could cut down the opportunity for immortality or a very long life for some whose life is quickly smothered out—though medical miracles are increasingly saving people who would otherwise die. Now, the potential for immortality or longevity has never been greater.

Given this possibility, more and more articles are appearing about recent developments, and a community of people interested in immortality has begun to form. So far, at least six organizations and foundations sponsoring immortality research are aiming at massive human life extension. These include the following:

- The Glenn Foundation for Medical Research, started by venture capitalist Paul F. Glenn in 1965, right around the time that the cryonics movement was born, has given grants of $60,000 to independent researchers who are doing research on aging. It works on promoting antiaging initiatives in large institutions, beginning with Harvard, and after that MIT, the Sal Institute, and Mayo Clinic. The foundation contributes over $1 million a year toward grants offered by the American Federation for Aging Research, a charitable foundation that supports research on age-related diseases.[5]
- The Ellison Medical Foundation, founded in 1997 by Larry Ellison, provides hundreds of thousands of dollars in grants to scholars doing research on remedies for aging, and it works closely with the Glenn Foundation in contributing these grants. So far, their efforts have paid off in extending the life of lab rats; and hopefully, the results of this research can be extrapolated to help humans combat aging in the future.
- The 2045 Initiative, led by Dmitry Itskov, aims to transfer one's consciousness to an artificial avatar-like body.
- Calico, led by Ray Kurzweil, seeks first to develop artificial intelligence and then work on major life extension technologies.
- The Methuselah Foundation, cofounded by Aubrey de Gray, created the nonprofit's main research initiative, Strategies for Engineered Negligible Senescence (SENS), with a goal of finding drugs that cure seven types of age-related damage to the cells on the theory that cell rejuvenation can combat and reverse aging: "a loss of cells, excessive cell division, inadequate cell death, garbage inside the cell, garbage outside the cell, mutations in the mitochondria, and crosslinking of the extracellular matrix." An underlying premise is that the human body is essentially a machine with a structure that determines its functions but that could fall apart at any time. So, if it is possible to restore the bodies at the molecular and cellular level, that will restore the body's functions too, thereby rejuvenating the body.
- The Alcor Life Extension Foundation, which calls itself the leading life extension organization since 1972, employs cryonics freezing of the head or full body. The goal is to bring people back to life, when the technology can revive the body without cellular damage.

Funding has now been offered for new research to bring the potential for living forever even closer, such as a $1 million Palo Alto Longevity Prize offered by Silicon Valley physician and hedge fund manager Joon Yung. His belief is that humans can live 1,000 years or more, and he wants scientists to "hack the code of life" to increase the life span of mice by 50% or more, as a way to start somewhere.[6] Another recent grant is a $5 million interdisciplinary grant sponsored by the John Templeton Foundation from 2012 to 2015, called "The Immortality Project." This grant has funded 34 research projects by researchers throughout the globe and has begun to compile the results of science, philosophy, and theology projects.[7]

The Scientific Search for Immortality

In response to this growing interest by scientists in the potential for immortality, *The Science of Living Longer* is designed to look at these modern developments, which offer the potential for both a much extended life and some form of actual immortality. The underlying premise of these efforts is that the basis for individual consciousness and identity lies in the brain. This is what can endure, whether one thinks of this essence of one's personality as conscious awareness, spirit, or soul, which lives within the neurons and synapsis of the brain.

Given this premise, the modern scientific search takes various forms to be discussed in this book, including preserving the head or body after death until it can be restored to life, revitalizing the cells to stop or reverse the aging process, providing the individual with newer body parts so that he or she becomes something of a cyborg, and even downloading or copying the brain so it exists in a computer or in the cloud, based on the view that the brain operates much like the hard drive of a computer. Once any of these models become possible so that a much longer life or immortality is possible, there are all sorts of social effects to be considered, such as the potential for overpopulation, the loss of jobs for the younger members of society, and new political power adjustments in a society increasingly dominated by long-life individuals, who have a much longer time to acquire skills and build relationships with others living much longer lives.

The Science of Living Longer addresses all of these issues. More specifically it covers these topics:

- the early scientific efforts to create the hope of immortality through cryonics beginning in the 1950s
- the major streams of scientific research and technological developments on immortality and life extension
- the role of Silicon Valley billionaires and venture capitalists in funding much of the current research

- the rise of immortality and life extension organizations and communities
- the development of antiaging and health products and activities
- the effects of living longer lives or living forever on society
- what individuals might do to increase their prospects for living longer or forever

The book is divided into three parts. Part I deals with the different types of scientific and technological developments that occurred and the results of this research into living a much extended life or living forever. It includes a discussion of the rise of different groups and organizations interested in the search for immortality and longevity. Part II looks at the impact of these developments on society and what individuals can do to improve their own chances for living much longer, healthier lives and perhaps even forever. Part III examines some of the techniques for preserving and transferring the brain to other bodies, both human and mechanical, and explores the latest bio-tech approaches for living longer.

PART I

The Search for Immortality

CHAPTER ONE

Strategies for Living Longer to Increase the Chance of Living Forever

Today, the growing boom in efforts to increase healthy longevity, as well as the potential for living forever, has been fueled by the aging population and the growing biomedical and technology boom that is in turn fueled by venture capital. Much of this funding has come from aging high-tech billionaires who are supporting research into biomedical and high-tech solutions for reversing the effects of aging and providing new ways to preserve one's consciousness despite any physical decline. Even problems of overpopulation, mass immigration, and the prospects for travel to other planets play into this booming interest in life extension, since it becomes possible to transcend the modern turmoil on this planet by spreading the population to other planets and galaxies or perhaps humans might, as some technologists put it, become a new evolved species that might escape the boundaries of normal human bodies and exist in a kind of Internet cloud.

To read research reports and the growing number of articles and books on the potential for immortality is to look at what may seem like science fiction. It is like entering another reality, where all kinds of scientific and technological leaps are possible, based on cutting-edge research on stem cells, cell rejuvenation, robotics, exoskeletons, and extending the lives of worms and mice. Underlying all of these streams of research and conjecture is the belief that immortality is possible, maybe in the next decade or perhaps in 30, 40, or 50 years, so that those living today might still enjoy eternal life. They just have to live long enough to enjoy the fruits of the new research in biology and technology, or perhaps, if they go the cryonics route, they can be revived through new thawing technologies that leave their brain and other cells undamaged, so their full bodies or heads transplanted onto new bodies can later be revived. As improbable as this may sound, those who are part of this new scientific research or part of the scientific community of believers insist all of this will soon be real for modern science and technology, with its godlike superpowers, will make it so. It just needs to unlock the door to immortality that resides within the human imagination, if not within the cells themselves.

True, many doubters are still writing articles that dispute the ability of scientists to successfully prolong the life of humans by using the same techniques that have been used to prolong the life of earthworms or mice, because these methods have not yet been applied to humans. Many suggest that the human brain is too complex to make this possible. Or sometimes the critics pounce on a particular stream of research, such as the studies of cell rejuvenation, which suggest that aging can be reversed, to claim the research methods or results are flawed. Yet supporters of this possibility of living forever largely ignore these critics, since they prefer to look to the possibilities of the future. As a result, the thrust of their research reports is mostly positive and celebratory, as one might expect from pioneers pressing forward into a new land—in this case, a world with the possibility of immortality.

At the same time, a growing community of people—from those in the rapidly expanding population aged over 70 to middle-aged baby boomers—is especially interested in supporting the positive results of these researches. After all, it is to the benefit of these people if they can live longer, healthier lives, if not forever, so more and more of these individuals are becoming active supporters of these different research strategies.

In effect, these different strategies are like different pathways to immortality—or at least to the potential for living a much longer, fuller life. And it is always possible to select multiple paths, much like one might have a portfolio of stocks, bonds, and other financial investments, given the uncertainty about which instruments will be the best performers. Or to take another well-worn comparison, one might do better to have his or her eggs in different baskets than to put them all in the same place. Likewise, these different strategies for achieving immortality are like holding stocks for the future or multiple eggs, unsure which ones will actually hatch.

Thus, the following approaches to immortality and longevity should be seen in this light—one or more of the approaches may work or may not.

In the meantime, billionaire investors and millions of believers are supporting different possibilities, believing we really can live forever, whether in bodily form or by transcending our human bodies and becoming some other form of continued consciousness.

What are these different approaches? How do they work, and who believes in them? Let the games begin.

Live a Healthy Lifestyle until the Later Approaches Work

One way to think of the potential for living forever is as a series of bridges to the new future biomedical and other technologies that are expected to be available in 15–25 years. For now, they aren't yet ready for prime time, so to speak; but if you can hold out long enough, they will be, and, presumably, future generations will benefit from these new technologies. As Ray Kurzweil and Terry Grossman describe in "Bridges to Life," a chapter in the massive tome *Futures of Aging*, the first bridge comprises present-day therapies and guidance that "should enable many people to remain healthy long enough to take advantage of the second bridge—the Biotechnology Revolution, which is providing us with the means to reprogram the outdated software that underlies biology."[1] Through reprogramming, scientists will be able to change the genetic and protein codes that make up our biology, so we can prevent diseases and aging. Then this change will prepare the way for the third bridge, which they describe as the "Nanotechnology-AI (Artificial Intelligence) Revolution." This final step will enable humans to make changes at the molecular level to build our bodies and brains. While Kurzweil and Grossman don't mention such strategies as freezing the head or body to revive in the future, using organ transplants or robotics to create a new, revitalized physical body or transferring the consciousness into other bodies or forms, their bridge metaphor is a good one, in that the first bridge leads to many future bridges, which can branch out or extend from one another.

Perhaps another good metaphor is that of the human tree of evolution, leading from an apelike ancestor, where branches in one direction led to modern-day apes that are close relatives, like the gorilla, chimpanzee, bonobo, and orangutan and branches in another direction led in "fits and starts" to modern humans. Along the way, starting with the *Australopithecus afarensis* and *africanus* line, there were some dead ends, such as *Homo rudolfensis*, *Homo ergaster*, and *Homo heidelbergensis*. But a line of steady progress went via *Homo habilis* to *Homo erectus*, and eventually to *Homo sapiens*, and according to some trees, there is now a *Homo sapiens sapiens*, a subspecies to which all humans belong. And perhaps as these new technologies develop, there may be a further subspecies that will reflect the future bioengineering that may occur, such as *Homo sapiens biotechno*, a name I have coined, to reflect this evolutionary transformation in the future.

But then, that is only a possible future. This is now, and this is what individuals can do in the present to get to the next bridge, according to Kurzweil and Grossman. These are probably things you have already heard about from health and wellness practitioners, nutritionists, and fitness coaches to make yourselves healthier and fitter today. But these strategies can now be marshaled for a larger purpose of not only improving your current well-being but also helping you get through the next decade or two so that you can take advantage of these different types of programs enabling you to live much longer and maybe forever. As Kurzweil and Grossman note, "We are only about 15 years from the maturation of the second bridge, and about 20–25 years from the full realization of the third bridge."[2] So if you are reading this now, you may well be able to cross that bridge when you come to it. By putting into practice these current strategies to live a healthier lifestyle today, you can gain access to the future strategies of tomorrow.

The four major current strategies include calorie restriction, exercise, nutritional supplements, and using predictive genomics to learn about likely future diseases, so you can begin to take preventive measures now.

Restricting Your Calories

Restricting calories may be a familiar admonition for anyone seeking to keep his or her weight down and stay physically fit, but researchers have also found this restriction to be related to increasing longevity. The basic requirement for calorie restriction, also known as CR, is to eat less food but still obtain sufficient nutrition, either through food or through nutritional supplements. This method does not work if you don't eat and thus starve yourself, which can lead to organ shutdown and seriously damage your health or even lead to death.

The effectiveness of calorie restriction has been widely researched in numerous species, including *Drosophila melanogaster*, or the common fruit fly, and mice. As reported by Mark A. Lane and his colleagues,[3] over 2,000 animal studies with different species showed dramatic results in increasing average life expectancy. Thus, there is extensive evidence, as Kurzweil and Grossman concluded, that "restricting calories slows down aging and can extend youthfulness,"[4] at least in animals. Additionally, some studies of humans practicing calorie restriction have shown a reduction in indicators of disease and aging. One reason for these favorable results is that animals with restricted calories—and presumably this finding would apply to humans too—have a lower level of free radicals, a chemical that causes a gradual deterioration of body tissues, especially the cell membranes. So these animals have less damage from free radicals to their cell membranes. Moreover, they have stronger DNA-repairing enzymes, which are important for staying healthy, since random mutations in the DNA can result in cancer and speed up other aging processes. But with more effective DNA-repairing enzymes, aging is slower and these animals have fewer tumors. Even though all animals on calorie-restricted diets do eventually die, in general, researchers have found that the optimal diet restriction is for animals to eat about two-thirds of the calories they would eat normally, along with adequate nutrients.

While controlled lab tests are not possible with humans, some studies have shown the benefits of living with a calorie-restricted diet for humans. For example, people living in the Okinawa region of Japan compared to those on the mainland have 40 times the number of centenarians, experience little serious disease before they are 60, and remain active longer, apparently due to their lower caloric intake. Some researchers have even suggested that the human life span might be extended to 180 years, based on the results of animal studies. Thus, as Kurzweil and Grossman conclude, even though the true quantitative benefits of restricting calories for life extension in humans is unknown so far, we do know that "restricting calories can improve human health and reduce many risk factors for life-limiting diseases in people."[5] However, the extent of this benefit depends on when you start to restrict your calories, since it only applies to one's remaining life expectancy. Thus, the sooner one starts restricting calories, the greater the benefit.

How much should you restrict them? While the reduction is typically 35 percent in the animal studies, Kurzweil and Grossman suggest consuming 20 percent fewer calories, along with following certain guidelines that include eating[6]:

- foods low in caloric density, such as eating oranges rather than drinking orange juice, and eating low-starch vegetables such as broccoli, instead of potatoes or rice
- foods rich in fiber, since these provide bulk and texture with no digestible calories, plus they lower cholesterol levels and reduce the risk of colon cancer
- starch or fat blockers, such as Precose, which delay starch absorption, and Orlistat, which blocks up to one-third of fat calories

In addition, it's best to avoid foods that are high in simple sugars and starches, and eat a diet that is low in most fats, while emphasizing healthy fats, such as those in fish oil, nuts, and olive oil. Then, too, Kurzweil and Grossman recommend certain substances that contribute to restricting calories, such as the prescription drug metformin, and resveratrol, which is found in red wine and perhaps is a contributing factor to health among the French, who regularly drink red wine with their meals.

Getting Your Exercise

Another popular admonition for better health also turns out to be a good practice for living a longer life—getting regular exercise. This can either be aerobic exercise, sometimes called "cardio" exercise, which involves stimulating the heart and breathing rate by pumping oxygenated blood through the heart to deliver oxygen to working muscles, or strength and resistance training, such as lifting heavy weights to increase muscle strength. As Kurzweil and Grossman point out, regular exercise contributes to longer living, since "all major progressive diseases, such as heart disease, stroke, and type 2 diabetes are dramatically reduced by regular exercise."[7] Presumably, exercise is so important because of our heritage from hunter-gatherers many, many centuries ago. This lifestyle involved extensive exercise as individuals tracked down game or gathered food, in contrast to a new factor contributing to modern degenerative diseases— our sedentary lifestyle, where most of us sit at desks at work, hunch down at computers, or spend hours in front of a TV set or at the cinema.

In turn, research supports this development. For example, in one eight-year study published in the *Journal of the American Medical Association*, S. N. Blair and his associates[8] examined the relationship between the fitness level of 13,344 participants and their mortality rate. They divided the participants into five categories based on how much they exercised, from being sedentary with no regular exercise program to those who attained a high level of fitness by walking or running 20–30 miles a week or more. As they found, people in the moderate exercise group had a 60 percent lower death rate than those who were sedentary, and those in the high-fitness group had an even lower death rate.

Thus, besides reducing calories, you should regularly exercise in three ways, according to Kurzweil and Grossman as well as many health and fitness coaches and trainers. In fact, many gyms and sports centers include equipment for all of these exercises and thus have classes with trainers, in which participants can work out with others, often to music, so they can learn to do it right. One type of exercise is aerobic exercise, such as walking, bike riding, swimming, jogging, and routines like Zumba, which reduce your risk of heart disease, cancer, and other diseases. Plus, these exercises can help you lose weight, lower your blood pressure, reduce your cholesterol, help you sleep better, and improve your mood.[9] The second type of exercise is resistance or strength exercise to increase the strength of your muscles. The third type is stretching, so you increase the extent of motion in your joints, since muscles, tendons, and ligaments tend to shorten due to aging and after performing other exercises, but with flexibility training or stretching, you can slow down this process. With all these exercises, it's important to participate in them regularly, ideally for at least 20–30 minutes a day, two or three times a week on nonconsecutive days. If you participate in a gym or work with a trainer, you will get guidelines on the best ways to work out based on your level of fitness, age, and other factors, when you first start exercising.

Adding Nutritional Supplements to Your Diet

The third rail in living a longer healthier life is adding vitamin and mineral supplements to your diet. Since most individuals need them, even if they have a good, healthy diet, a billion-dollar global industry has grown up to provide all kinds of supplements in your local supermarket, with different companies claiming their formula has the best combination of ingredients. Often these companies claim extensive supporting research, sometimes with secret or hard-to-get ingredients, and many companies have an army of recruits, usually women, who pitch these products in various ways, from online sales to attending business-networking meetings and trade shows and reaching out to friends and neighbors. Among the most well known of these companies are Amway, Herbalife, Shaklee, Arbonne, Plexus, Melaleuca, and Juice Plus.

These companies are based on the premise that your diet doesn't provide enough vitamins or minerals, so you need to supplement them. Why? One of the reasons is that current farming methods provide lower levels of these nutrients. Another claim is that very few people eat enough fresh produce to get enough nutrients without adding these supplements. Then, too, as people age, they obtain fewer nutrients from their food, because their digestive functions are reduced, and some recent research suggests that many people have genetic defects, requiring them to compensate by taking the needed nutritional supplements. When you take them, you may not notice much difference, as was my experience, even though representatives for these companies report extensive testimonials from happy customers who experience great results, such as having more energy and fewer illnesses, and getting over colds and the flu more quickly. Many report their own healing experience, which often leads them to become a sales rep for the company in order to help others have the same transformative experience and to make money.

So should you take a nutritional supplement? Aside from the recommendations from company reps, who usually advise that you get a package of supplements totaling a $100 or more each month, Kurzweil and Grossman recommend that almost everyone over 30 take both a multivitamin/mineral formulation that meets the needs for optimal nutrition and fish oil. These daily needs are sometimes greater than the Recommended Daily Amounts, which are largely designed to prevent deficiency diseases like scurvy and rickets. Generally, Kurzweil and Grossman advise that you need more than a once-daily multivitamin pill. Expect to supplement that with vitamins B12, C, D3, E as well as other commonly recommended nutrients—it might be good to check with a nutritionist or other health professional to determine what combination of vitamins you need. Additionally, eating fish several times a week or taking a fish oil supplement is recommended, since fish is considered a good source of the omega-3 fatty acids, which contribute to reducing any inflammations in the body. Again, check with a nutritionist or other health professional to determine what you need.

Learning about Yourself through Predictive Genomics

The fourth way to create a health program to hold off aging is by learning about your genes, to find out about your health risks. A genetic analysis won't tell you if you will get a particular disease, but you can learn about your tendencies; then your lifestyle choices and environmental conditions play a great part in affecting whether you get certain diseases or not. The advantage of learning about yourself through predictive genetics is that a doctor can provide you with individualized therapies based on your genetic profile, instead of applying the same general medical practices to everyone. In turn, this knowledge of your genome can help you determine your optimum type of diet, exercise program, and nutritional supplements.

According to geneticists, there are about 23,000 human genes, and the analysis of these genes has just begun. But you can still find out in minutes about yourself, instead of spending weeks going through individual tests for different conditions. Numerous genomic tests are now available commercially, and these can tell you about the degree to which you are predisposed to many serious diseases. These can be prevented or modified to be less serious, if you have advance knowledge so you can prepare in advance. For example, you can learn about your predisposition to heart diseases, Alzheimer's, type 2 diabetes, or cancer.[10] Increasingly, more information is becoming available through testing, and the costs have come down substantially. For example, you can obtain genetic testing through 23andMe, Ancestry, and many others, for as little as $99—and you can learn more about your own ancestry too.

The basic way this genetic testing works is that it attempts to identify the most significant of the up to one million single nucleotide polymorphisms (SNPs) that each person carries. Once you know this information, the test can determine how likely you are to be predisposed to getting a particular disease or experiencing a certain health risk under particular environmental conditions or lifestyle choices. At the time Kurzweil and Grossman were writing, these tests were conducted for individual genes. Now these companies can analyze most or all of your genomes for a much lower cost than when genetic testing began and cost about $30 per gene for about three dozen SNPs.

Needless to say, once you know about the hundreds of SNPs that can be modified by making changes in your diet, lifestyle, nutritional supplements, or prescription drugs, you can work on making these changes. Preferably work with an expert on genetics testing to decide what is best for you.

Packaging a Perfect Diet for Living Longer

Given this formula for living a longer, healthier lifestyle to increase longevity, if not achieve immortality, it is not surprising that a growing number of companies are packaging parts of this formula to help individuals work toward this goal. They have also been developing consumer literature, drawing on longevity research with humans and animals, along with bottled liquids and capsules, marketing materials, websites, and other aids, to promote sales to a growing market for the immortality dream.

For example, while I have been researching and writing this section, I got a book called *Immortality* by Joel D. Wallach and Ma Lan, which was accompanied by a CD for a calorie-reduction diet and a threefold color brochure announcing a Rebound Fx sports drink available through Mineral Wellness (www.mineralwellness.com), one of a network of distributors for Youngevity (www.youngevity.com), based in Utah. When I went to the website and checked on the available products, there were dozens of individual bottles and nutrition bars with prices ranging from about $50 to $125, ready to be shipped to countries around the world, and $500 multipacks priced at $499 for more consumer appeal. The site also announced a conference in Texas in August 2017 plus a line of beauty products and energy drinks.

Immortality focuses on the two legs of the four strategies for living longer: following a calorie-restricted diet without malnutrition and adding nutritional supplements, even though their research has led them to claim that everyone needs 90 key nutritional plant minerals missing from most of the foods we eat today. While their focus is on the secret of becoming a centenarian, the title *Immortality* suggests that living even longer is possible by following their recommended diet, based on their studies throughout the world. The book, in turn, is like a bible for their longevity empire. As they write in the introduction:

> The secret of becoming a centenarian is "hidden" in plain view! The epiphany revealed in IMMORTALITY, is a compilation of the life's work (94 years between the two of us) of two physicians from separate worlds—China and Australia . . .
>
> In IMMORTALITY, the two of us view health, life longevity, disease and death through the eyes and understanding of a farmer, animal and human nutrition, soil and plant nutrition and chemistry, veterinarian, primary care physicians, surgeon, Chinese Traditional Medicine and pathologist—this is the first time in history and human endeavor that these crafts, trades and professionals have come together under one roof to solve the mysteries of the "centenarian club" and uncover the universal currency of the age beaters.[11]

But beyond promoting calorie restrictions, Wallach and Lan focus on our lack of plant minerals, due to the shift from wood as a universal fuel to electricity caused by the social change that occurred beginning with the Industrial Revolution. The result is that while we have benefited from the new technologies that have developed since then, the unintended consequence has been the loss of these plant minerals, so now humans have to supplement them with these essential minerals. As Wallach and Lan explain:

> We no longer have easy or unconscious access to traditional sources of plant minerals required for healthy human and animal life and as a result we must supplement with all of the known essential minerals . . .
>
> The missing ingredient, plant minerals, was at one time in history a universal and basic by-product of daily life. We burned wood and other carbon fuels for heat, cooking, and light . . .

You cannot effectively use any nutrients, phytonutrients or raw materials physiologically without mineral co-factors—even oxygen and water are unavailable to you at the cellular level without mineral cofactors![12]

Reportedly Wallach and Lan discovered this truth through their visits around the world, where they found that the longest-living people lived in remote and poor environments. The reason they lived so long is that they unconsciously submitted to calorie-restricted diets of about 1,200–1,500 calories per day, consumed large quantities of antioxidant foods and drinks (such as coffee, green tea, chocolate, wines, fruit, sweet potatoes, kelp, and vegetables), and consciously and proactively used multiple sources of minerals (including flood silt, dust compost, and plant mineral supplements, such as the ash from wood, kelp, peat, and manure) in their foods and in their gardens for fertilizer.

As a result, Wallach and Lan conclude that anyone wanting to maximize their chance of becoming a centenarian in our modern high-tech world needs to "consume a Calorie Restricted diet, consciously supplement with plant minerals, as wood ash . . . is no longer commonly available to fertilize our gardens or add to our food . . . and consume large quantities of high grade antioxidants each day."[13]

The rest of the book is devoted to providing research support for the need for a calorie-restricted diet and mineral nutrients, and describing each of the nutrients at length. For example, Wallach and Lan describe how adding nutrients to pelletized rations resulted in eliminating by prevention or curing as many as 900 different diseases in laboratory animals, and this process also doubled or quadrupled the life spans of all animal species in the home, farm, lab, and zoological park. They point out that these "gerontological miracles" came about because adding "vitamins, minerals, trace minerals, rare earths, essential fatty acids, amino acids and antioxidants to a commercially prepared diet of alfalfa pellets or dog food" made the big difference. They emphasize that the three methods of increasing the maximum life span in laboratory animals fed "perfect diets" are having a lower metabolic rate through hypothermia, having a calorie-restricted intake without malnutrition, and enhancing their antioxidant protection through genetic engineering and supplements.[14]

They additionally give some examples of well-known people who lived longer than usual lives over the centuries, such as Luigi Cornaro, famous Renaissance author of *The Art of Living*, who lived 103 years (1464–1567) at a time when the normal life span was about half what it is today. Even though Cornaro lived a dissolute life filled with gluttony, leading to disabling health conditions by the time he was 37, including suffering from gout, stomach fever, arthritis, and morbid obesity, he turned everything around by becoming his own physician and finding out from the old women working in the fields what they ate. As a result, he became aware of the peasants' calorie-restricted diet, since they were heavily taxed and could only obtain food from their small plots of land. Soon he changed his diet accordingly for the rest of his life, and he became lean and healthy. Wallach and Lan similarly cite the success of three European physicians—Sir Hermann Weber (1823–1918), Sir James Crichton-Browne (1840–1938), and Dr. Alexandre Guenoit (1832–1935)—who combined walking and different types of exercise with a healthy diet.[15]

Why does the calorie-restricted diet without malnutrition work so well? Because it leads to several key changes in the body that slow aging and keep the individual feeling more fit and healthy. Among them are these seven key effects, shown by research on laboratory animals fed a low-calorie, high-nutrition diet. This calorie-restricted diet does the following:

- lowers circulating glucose and insulin levels
- lowers the percentage of body fat
- lowers blood and tissue levels of growth stimulation
- reduces cell loss from the aging process
- decreases the effects on the body of inflammation or free-radical injury
- encourages a youth physiology—and it has the most effect when started in infancy, although longevity is still increased dramatically when the diet is begun in mid adulthood

Additionally, some researchers found that intermittent periods of fasting for rats, such as every other day, every third day, and every fourth day contributed to a 20–30 percent increase in the maximum life span from 800 to 1,100 days.[16]

While the research that Wallach and Lan cite comes from studies of laboratory animals, the inference is that this will similarly have a dramatic effect in increasing the life span of humans who follow a calorie-restricted/high-nutrient regime.

They similarly argue for the benefits of reducing or removing free radicals through antioxidant enzymes, since free radicals trigger abnormal reactions in the cells that are energetically wasteful and irreversible. The reason for this is that these free radicals possess an extra electrical charge or have a free electron that sets off abnormal reactions at the cellular level, and most of the free radicals in the body are oxidants. Thus they urge that one should use antioxidants to avoid the toxic antiaging effect of these oxidants.[17]

The remainder of the book is devoted to describing in detail the essential nutrients and needed minerals, followed by examples of "age beaters" around the world and their practices and diets.

Since I'm not a biologist, nutritionist, or other scientist able to critique the specifics of this antiaging program, I can't critique the underlying science supporting a certain type of diet to promote longevity, which might lead to immortality if you live long enough to benefit from the expected medical and technological revolution. But there does seem to be a growing consensus among researchers, along with a rapidly growing market for commercializing antiaging products. At one time, health and medical professionals emphasized that better eating, along with exercise, would contribute to a healthier, longer, fuller, and happier life. But now it would seem that these admonitions and products designed to promote good health have been mobilized to support the goal of living longer.

From the perspective of marketers and entrepreneurs, this shift has created new markets for achieving the modern-day "fountain of youth," and the programs and products developed by Wallach and Lan are a part of this growing market. From another perspective, this new emphasis on living longer is providing a major contribution to the medical and technological developments that might make even further longevity and real immortality possible, by prolonging the lives of those who might take advantage of the new medical and technological discoveries when they come online in 15, 20, 25, or more years.

As will be discussed further in the next chapters, the only one of these technologies that people can use today is cryogenics, based on freezing their heads or bodies now in the hope that they can be revived sometime in the future, when they can be brought back so that their brains (including their intellect and memories) are still operational and they maintain their old identity. It's not certain when or whether this technology will work, even though individuals today can opt to have this cryogenics suspension once they die, with the hope of a future life, whereas for others, death is still final now, even though in the future, immortality must be possible if the living can only keep living long enough to take advantage of the immortality revolution that lies ahead.

What are these different technologies and how do they work? The next chapter will describe the different paths possible for immortality for those who are living now.

CHAPTER TWO

The Different Paths to Immortality

Today the search for immortality has gone in a number of directions based on biogenetic research and technological developments. With the exception of cryonics, which is a "try now and hope for the best" strategy, these other methods provide opportunities for life extension now, such as through cell rejuvenation or replacement organs. But the prospects for true immortality, if possible, are off in the future, such as research on brain transplants or mind-to-machine brain downloading. But any real prospect of their success is from 15 to 25 years or even longer in the future, depending on the rate of progress in that field. In the meantime, you can apply some of the strategies for living longer through better health, nutrition, and exercise described in Chapter 1. Then maybe you might live long enough for any of these strategies for immortality to work for you.

Here I want to introduce these different paths and describe them in more depth in future chapters. These paths fall into two basic categories by developments in biotechnology and technology, including miniaturization, nanotechnology, and artificial intelligence. Plus there are some spiritual beliefs about the immortality of the soul as it moves from body to body.

The Biotechnology Paths to Immortality

Some of the major biotechnical pathways involve changing on a cellular level or combining biological and technological developments.

Cyronics

This is one of the earliest paths to immortality that began with cryonics' promise for the future in the 1960s. In fact, I was there in its early days, when I attended hopeful meetings in New York City in 1968. The meeting leaders described cryonics as a new science breakthrough, and now they were inviting participants to enjoy this opportunity to freeze themselves upon their death with a chance to be unfrozen and revived. It was, as they explained, a chance of a lifetime for a renewed life.

Basically this approach involves freezing the body in liquid nitrogen after replacing the blood with a cryoprotectant fluid or gel to keep ice from forming in the cells as the temperature is set to a very low level—about −196°C. Then the body is held upside down so that the head is kept at the lowest temperature. While some bodies might be placed in a single tank, others might be put in a giant steel tank that can hold up to four people. Then the frozen person can wait—and wait—and maybe wait some more, until the technology can not only revive the body without damaging the cells but also medical science has advanced enough to cure the underlying disease, disorder, or injury that caused the person's death. If a person can't afford freezing his own body, only the head will be frozen to be reattached to another body sometime in the future.

The process may sound a little like taking your car to the local car wash and coming out with a much newer, spiffier, faster car. In any case, as of 2013 there were over 250 people who were preserved in this way and 1,000 who had signed up to be frozen. To do so, they or their family had paid a hefty price tag—about $50,000–$200,000—to be frozen and to have regular maintenance over the years to preserve one's body or head in an optimally frozen state to avoid meltdowns that would wash away one's dreams of a future revival.[1]

Curing Diseases of Old Age

Some efforts to attack the diseases of old age involve fixing the cells or even performing fixes at the molecular level; other methods include discovering hormones and drugs that promise a longer, or forever, life. Some researchers have also been studying the oldest supercentenarians, who live beyond 110, to see how they have done it. As individuals are experiencing longer, healthier lives today due to better nutrition, health practices, and medical treatment, their numbers are increasing. According to "Will Technology Help Us Live Forever?" by Madhumita Murgia, as of 2013 there were about 316,000 living people over 100 and 82 individuals over 110. By 2050 over three million people are expected to live to over 100 due to advances in medical technology.[2]

Unfortunately, as the life span increases, people are increasingly experiencing the chronic diseases of aging, such as cancer and Alzheimer's, due to cells aging and the immune system being less resistant. A big research push has gone to finding ways to treat these different diseases and the genetic factors that lead to aging. To this end, a number of venture capitalists (including Brian Singerman, a partner in the Founders Fund, and Peter Thiel and Sean Parker, both billionaire tech investors) have been investing in biotech companies that are seeking ways to cure cancer and all viral diseases. They have high hopes this approach will work, since in Singerman's view, within 10 years, we will be able to cure all viral diseases, and we will more clearly understand aging, what causes it, and how to stop it.[3]

One promising approach has been studying the hormones and drugs that can reverse cellular death. Initially researchers study the effect of these drugs on mice before they test and use these drugs on humans. So far researchers have found several drugs that have increased mice longevity. One is an organ transplant drug called rapamacin, which extended the life span of the mice in the study by 25 percent and protected them from getting cancer. Another promising drug is resveratrol, a molecule in red wine, which has a positive effect on cell metabolism.[4]

Fixing the Telomeres

Another strategy, which dates back to the 1980s, is to fix the telomeres, which are the caps of repetitive DNA at the end of the chromosomes. These telomeres act like putting on boots to protect your feet from the cold and snow, because when a cell divides, the DNA is not replicated perfectly at the ends, so the strands of DNA, also called chromosomes, get shorter each time the cell divides. But the telomeres act like protectors, so the actual DNA doesn't get shortened during this replication process.[5]

Unfortunately, as scientists have found, each time a cell divides, these telomere caps shorten, until they are shortened too many times and eventually stop dividing and die, which brings about aging, because without these buffers, the cells lose their necessary DNA, so they begin aging.

But some scientists are working on the theory that the telomere enzyme in these cells can be reawakened in dying cells. That will stop the telomores from getting shorter so that human aging can be slowed, stopped, or even reversed, making human immortality possible. As proof of this theory, scientists have been able to replace the enzyme in old, declining mice, resulting in them becoming healthy again.[6]

Some of this research on mice has been conducted at the MD Anderson Cancer Center in Houston, where Dr. Ronald dePinho has been experimenting with techniques to keep mice from growing old. DePinho looked at the way in which chromosomes fray due to the shortening of the telomere, based on the notion that the fraying chromosomes might be causing some of the physical effects of aging. After he and his team found a way to turn the level of telomerase output on and off in genetically engineered mice, they discovered that the mice aged prematurely when there was no telomerase, so they were like 90-year-old humans, since they had shrunken brains, impaired cognition, thin bones, hair loss, and were infertile.[7] But once the researchers turned on the telomerase again, the organs started to restore themselves, cognition improved, the mice were once more fertile, and their hair was revitalized. Not only did giving the mice telomerase stop the gaining process but the animals seemed to become younger.[8]

However, the one problem with this line of research is that telomerase is linked to preventing and causing cancer. When the cells become cancerous, the telomerase level increases. By living longer, one can be more likely to get cancer, resulting in one's death. In other words, by reversing the effects of aging by increasing the level of telomerase, one is also increasing the potential for cancer starting or growing. Even so, DePinho and other researchers hope to develop a telomerase therapy to reduce the incidence of cancer, along with increasing the telomerase and keeping the telomeres from shortening.

So far this research only involves mice, but the ultimate goal is that if this approach can work in mice, it can be used to work in humans too. But will it? Possibly, because researchers at the Stanford University School of Medicine have developed a new procedure to lengthen the telomeres artificially. In brief, this involves modifying the RNA in cells that carries instructions on the making of proteins in the cells. If it works, researchers believe this new procedure might not only increase life spans but also overcome a variety of diseases that affect thousands of people today.

Can scientists stop the telomeres from shortening in humans and will that contribute to helping humans live longer or forever? Stay tuned. . . .

Rejuvenating the Cells

Rejuvenating the cells is an approach advocated by Strategies for Engineered Negligible Senescence (SENS), which is designed to make the cells young again in order to eliminate any diseases caused by aging. The goal is to develop multiple therapies and prolong life by this means, overcoming different types of cell damage due to aging.

The rejuvenation theory was first developed by studying lobsters and hydras since they don't show any evidence of aging. Now the researchers have primarily been studying aging in mice, and they have successfully increased their life expectancy. In fact, one mouse proved to be something of a long-lived life phenomenon since he lived to 1,819 days, even though mice normally live less than a year. So again, the results are confined to mice, although researchers hope these discoveries will lead to rejuvenating cells in humans so that they too can have healthier, disease-free lives and live longer.[9]

Working with Stem Cells to Create Transplants and Further Gene Editing

Another development has been research on stem cells (HSCs) in order to use these for blood-cell or bone-marrow transplants to preserve a patient's life. One of the problems in the past has been the difficulty of finding stem cells that haven't turned into other types of cells in the body, such as cells that make up the liver or lung or that repair an injury. But in 2014 researchers at the Harvard Stem Cell Institute had a breakthrough and were able to turn mature blood cells in mice into new stem cells by reversing the process whereby stem cells turn into other mature cells. Though researchers haven't yet conducted tests on humans, they believe this discovery could be a major breakthrough: instead of having to find gene-matched donors to perform a transplant, doctors can use a patient's own cells to create.

Other stem cell research at the Salk Institute has proved promising: researchers there discovered a type of stem cell in certain parts of a developing embryo that can more easily be grown in the laboratory and used in gene editing. In the past scientists were unable to use the stem cells in later stages of the embryo's development. Again, the research results are all due to working on mouse embryos. Potentially these techniques could be applied to human stem cells, such as replacing parts of the body that have degenerated due to age with new stem cells. It's a little like giving an exhausted, aging athlete a super energy drink so that he suddenly feels more energized and young again.[10]

Using Nanobots for Repairing the Body

The nanobot approach involves developing microscopic machines to fix the cells in your body at the molecular level, thereby increasing your resilience to disease and aging. To do so, scientists inject a solution of tiny micro-organisms into your body to make these repairs. For example, nanobots might destroy cancer cells so that the sufferer becomes cancer free. So far these nanobots appear to work on specific tasks. For example, scientists have used microscopic nanofibers of synthetic molecules to repair the optic nerves in blind hamsters. The process is a little like applying a miniscule material like duct tape to bind up the wall of the nerve so that it will function again.

Besides using this technology to repair human cells, scientists could use these tiny bots to scan the entire human blood system and send this image to a holograph or other virtual reality system so that doctors can carefully examine the system to find any risks of breaks, such as a potential aneurysm in the brain. A nanodevice with data on toxins and pathogens could be used to recognize and destroy an invasive agent in order to make the human immune system even more effective. Nanotechnology could be used to remove certain substances such as lipofuscin, which interferes with cell function and thereby contributes to aging.

Even though these technologies are still speculative, nanobots have already helped to lengthen lives in the fight against cancer, and they could well prove to be the future of the medical industry.[11]

Studying the Genes for Things to Change or Fix

Another more basic research strategy has involved studying the genetics of centenarians and supercentenarians for markers of longevity. One company dedicated to doing this is the Human Longevity Institute, started by Craig Venter, the genetics pioneer who originally sequenced the human genome, and Peter Diamandis, an entrepreneur. The Institute's goal is to sequence the genomes of 1,000 people by 2020, including the genomes of the supercentenarians, to look for the genetic factors that lead to living longer. Eventually the Institute plans to create a genome-sequencing service that individuals can use to examine their own genes, much as they do with companies like 23andMe.[12]

Another approach is biochemist Cynthia Kenyon's search for ways to turn off "aging genes" through genetic modification, called "gene therapy." In experimenting on roundworms she found that when the DAF-2 gene was damaged, the worms' life spans more than doubled since they aged half as quickly. As a result, a 10-day-old mutated roundworm was more like a five-day-old normal worm.

In turn, Kenyon suggested that this approach could be applied to increase the human life span. She studied the biochemical, phenotypic, and genetic variations in a group of Ashkenazi Jewish centenarians, their offspring, and offspring who didn't come from a line of centenarians and found that a significant number of Ashkenazi Jews who lived to 100 years or older had DAF-2 mutations, and presumably they passed this mutation down to their offspring.

She and her associate concluded that the Jewish centenarians had a genetic modification that enabled them to be more likely to live a longer life, and they passed this on to their offspring. The researchers also concluded that this discovery of one gene that contributes to longevity suggests the possibility of discovering still other genes affecting the aging process.[13] This further opens up the possibility of manipulating these genes or inserting them into the chromosomes of other individuals, thereby increasing the potential for longevity. It is like finding a magic pill that can be given to others so that they can live longer.

Still another approach to turning off the genes has been developed by Keren Yizhak and her colleagues at Tel Aviv University's Blavatnik School of Computer Science. They came up with a computer algorithm that predicts which genes can be turned off to turn diseased cells into healthy ones. To do so, this "metabolic transformation algorithm" takes information about two metabolic states and predicts the environmental or genetic changes needed to go from one state to another. Intriguingly, Yizhak applied this algorithm to the lowly yeast cell to predict which genes could be turned off so that the expression of genes in old yeast would look like that of young yeast.[14] Yeast may seem an unlikely candidate for such a test since it is normally associated with baking bread, but Yizhak used it because much of yeast's DNA is preserved in humans.

Eventually, Yizhak and her colleagues found that turning off two yeast genes, GRE3 and ADH2, extended its life span 10 times. Even though it might seem a giant leap to go from yeast to humans, Yizhak applied the metabolic algorithm to humans and was able to identify a set of genes that could explain 40–70 percent of the differences between old and young genes in four different studies. Many of these genes in Yizhak's model have been found to extend the life span in yeast, worms, and mice, and these results might be used in developing drugs that could target genes to turn off in humans to enable a longer life. This line of research on turning off selected genes might also be used to find drugs to overcome disorders due to problems with metabolism, such as obesity, diabetes, neurodegenerative disorders, and cancer.

Besides fixing genes, this approach might be used to turn off genes that contribute to the diseases that lead to aging and death.

Reactivating Silenced Genes with Some Chemical Assistance—from Honeybees to Humans

Another research approach that seems promising is to reactivate genes that have been silenced or switched off, since this silence can contribute to aging. To this end, Stanislaw R. Burzynski, president of the Burzynski Clinic in Houston, Texas, began by looking at the theory of gene silencing, which explains that a key factor in aging is the reduced expression of numerous genes. According to this theory, only 10 percent of our genes are active in our young adult life, and the remaining genes are silenced or switched off, although they can get turned on again by the "epigenome," which activates and silences the genes without changing the gene's sequence. Basically what happens is that most of our genes are initially active as the embryo develops, but they begin to be blocked by certain chemicals in the body since they no longer need to be expressed. Our bodies reach their optimal level of active and inactive genes at age 25; after that, groups of genes are turned off due to aging. As these genes are silenced, the typical signs of aging occur, such as graying hair, hair loss, wrinkled skin, reduced immunity, reduced detoxification, and cancer.[15]

However, as Burzyski explains, certain types of chemicals (such as certain peptides, amino acid derivatives, and organic acids) can help turn the genes back on. His group looked at how certain chemical factors—PG (pehnylacetylglutamine), isoPG (phenylacetylisoglutamine), and PB (phenlybutyrate)—affected the life span of honeybees and then suggested how these results might apply to humans. Why honeybees? The ancient Egyptians recognized the positive effects of proper nutrition on life span. The ruling class lived much longer than did the general population, in part because of their specially selected diet and in remedies such as honey that they used to treat illnesses. Archeologists discovered this by doing a chemical analysis of archeological artifacts, which showed that almost 60 percent of 900 Egyptian remedies contained chemicals found in honey from honeybees.

They also found the bees to be a perfect model for studying the relationship between diet, aging, and genetics. Honeybees live in three different social groups (workers, drones, and queens) that have very different life spans. The queens live for around two years, drones live for 10 months, and worker bees live for only five to six weeks. The differences result from what they eat, since they all start off with the same genes: the larvae that become queens are fed royal jelly for their whole life, drones receive it for four weeks, and workers receive it for only three days. When the researchers analyzed royal jelly through a chromatographic analysis, they found it had a high concentration of PB.

Other researchers have likewise found that PB and PG extend the life of bees. For instance, Professor Jerzy Paleolog at the Agricultural University in Lublin, Poland, found that worker bees lived about five times longer than usual after getting a single injection of PB and PG.

Research shows that these chemical substances work for humans as well. For example, Victor Segalen, at the Department of Pharmacology in Bordeaux, France, found that PG and isoPG in a group of oral supplements, cosmetic cream, and lotion significantly reduced skin wrinkles in a trial with 22 healthy volunteers. Certain chemicals have been seen to have an antiaging effect on the skin of humans, and perhaps future studies will find still other chemicals with antiaging results that contribute to longevity.

Using Blood Transfusions to Provide Younger Blood

Though it may sound ghoulish, researchers are exploring whether blood transfusions from a younger person to an older person will reverse some of the conditions related to aging. After this transfusion research had some promising results with mice, neurology professor Tony Wyss-Corya at Stanford University began a trial with humans to see if a blood transfusion from young people to Alzheimer's patients can help them regain their cognitive abilities.[16]

This research with mice began in 1956, when gerontologist Clive M. McCay at Cornell University in New York linked the bloodstreams of young and old mice together by sewing their flanks together. While the young mouse was full of energy, the other was low energy and feeble. But once their bloodstreams were linked, the young mouse aged prematurely while the old mouse became healthier and young. For nearly five decades McGay's research was ignored, but in 2004, Amy Wagers, at Harvard University's Departments of Stem Cell and Regenerative Biology, repeated his young blood-old blood experiment and found it worked. Why? Because she discovered that a protein called GDF11 was common in the blood of young mice but rare in the blood of the older mice. As a result, when it was introduced into the blood of the older mice, it led to much of the reverse aging that occurred, since GDF11 keeps the stem cells active in the blood. But in aging, when the GDF11 levels drop, the stem cells involved in tissue renewal are less active so that any healing is slower and other effects of aging recur. But because injection of young blood with a high level of GDF11 leads the stem cells to become active again, the signs of aging in the older mice go away as they produce healthy, younger tissues and might result in longer life spans. So if it works with mice, why not humans?

This research on rejuvenating the blood has more possibilities for rejuvenating other organs and functions in the body. For example, this same team showed that the GDF11 protein can also rejuvenate muscles and the brain, and another team found that injecting plasma from young mice into old mice can boost learning.[17]

This line of research seems very promising for humans as well, since it suggests that humans might similarly experience slower aging and longer lives if their stem cells are rejuvenated through younger blood or injections of the GDF11 protein to rejuvenate their muscles and brain.

Developing Antiaging Drugs

Developing antiaging drugs is another stream of research by both academic researchers and researchers at pharmaceutical and health care companies. For example, researchers found that one compound called sirolimus (or rapamycin), once used to suppress the immune system to prevent organ rejection following organ transplants, could extend the life spans in yeasts, worms, and mice. Other researchers found that the drug everolmius, used to treat certain cancers by reversing the normal deterioration of the immune system due to age, might be used to stave off many of the diseases older people might otherwise be more likely to contract and die from and thereby prolong their lives. In other words, strengthening immune systems with this drug might contribute to longevity.

Scientists at UC Berkeley discovered a drug called the Alk5 kinase inhibitor, which helps revitalize brain and muscle tissues, based on insights from their tests on mice. It does so by limiting the release of TGF-betal, a chemical that reduces the stem cell's ability to repair the body. As people age, they produce more of this chemical. If less of this chemical is produced, people may be healthier in their old age, since their bodies are better able to repair any damages, and they are less likely to get age-related diseases such as Alzheimer's.[18]

In turn this growing interest in antiaging drugs is apt to result in researchers developing even more drugs to combat aging.

Creating Commercial Life Extending Products

Building on this biotechnology research to extend lives, some companies are seeking to use research results to fund new products that consumers can purchase. For example, Alphabet, Google's parent company, has invested $730 million in a new company called the California Life Company (Calico). So far Calico has partnered with universities and pharmaceutical companies to create a number of life extending products. This research has also involved looking for longevity-related gene markers in individuals living 100 or more years and finding treatments for some of the major age-related diseases, including Alzheimer's and Parkinson's.[19]

High-Tech Solutions

Another line of research has been on applying current knowledge about machines and technology to create replacement body parts when the original parts fail.

Using Cybernetics

The cybernetics approach is based on creating prosthetics that can be added to the body when the biological counterpart breaks down. Thousands of people (such as wounded military personnel and athletes) are using prosthetic limbs, and surgeons are using heart stents and kidney dialysis to perform bodily functions. Many replacements still depend on organ transplants (such as kidneys and livers from living or recently deceased donors), but increasingly more and more body parts and organs can be replaced by machine equivalents. The result is more opportunities for an individual to survive—although more and more like a cyborg than a human being.[20]

For example, scientists, working with Neurobridge Technology, put a chip in the brain of 23-year-old Ian Burkhart, enabling him to move his paralyzed hands with his thoughts. The chip worked by bypassing his brain's electrical system to go directly to his muscles. In another case, scientists replaced the right eye that Canadian filmmaker Rob Spence lost in a shotgun accident with a video camera that transmits what he sees to a computer.[21]

While these cybernetic technologies have been used to improve daily life for individuals, they show promise for increasing the life span by replacing worn-out body parts with new high-tech components, foreshadowing the possibility of transferring a human brain into a newly rebuilt human body.

Creating New Organs through Growing Them in the Lab or 3-D Printing

Another high-tech approach is growing organs in the lab or creating them through 3-D printing. As of this writing, these techniques have been used to print livers and kidneys, to turn skin cells into stem cells, and to turn stem cells into organs. Then these newly grown or printed organs can replace damaged ones.

The 3-D printing process, called "bioprinting," creates organs by print-ing live cells rather than paper, plastics, or other inorganic materials. Sci-entists first harvest human cells from biopsies or stem cells. These cells then multiply in a petri dish, and scientists then feed the resulting mix-ture, like a biological ink, into a 3-D printer, which turns them into a specific 3-D shape. These 3-D-printed cells are then placed in the body to integrate with existing tissues.[22]

The goal of this technology is to increase the human life span by replac-ing deteriorated organs so that the body remains young and healthy. The-oretically this means scientists can eventually reproduce every part of the body, including blood vessels, bones, skin, fat, and muscles. Perhaps this approach might be combined with advances in robots and prosthetics so that one becomes literally a new man or woman. The one part not accounted for through these technologies is the brain, although that might be rejuvenated by other procedures or reduplicated like a computer file, now being researched by still other scientists and technologists.

Using Cold Storage and Resuscitation Techniques to Lengthen Life

Still another recently developed procedure can literally bring someone back from the dead. In this procedure, called "cold saline resuscitation," a dying patient is first placed in a state of animation by replacing the patient's blood with an infusion of cold saline, which drops the body's temperature so that the patient goes into a state of hibernation. Once a patient is in this state, doctors are able to fix many things that could oth-erwise lead to the patient's death, such as critical injury, hemorrhage, or organ failure. With the patient in the cold storage state, the doctor can operate to fix damaged organs, and if cloned organs are available, the doc-tor can use these to replace a damaged organ.[23]

Brain Downloading and Uploading

Downloading or uploading the brain—sometimes known as "brain emulation"—is based on imagining the brain like a computer hard drive, where the mind is formed like data through bits and bytes. Presumably the human mind based in the brain consists of thoughts, beliefs, memo-ries, emotional responses, personality traits, and other qualities that cre-ate the unique individual. The theory is that scientists copy or download this data onto a robot or a computer or onto the Internet cloud. In this way the mind can live apart from the body.

Currently there are two proposed methods for making this process work. One is the "copy-and-transfer" method, in which an entire brain is scanned, and every region is comprehensively mapped, so that every electron is included in the mapping. That copy is then transferred onto a computer device. This device need not be a stationary desktop, laptop, or mobile unit; the copy of the mind could conceivably be transferred into a robot with prosthetic arms and legs, and this robot could even resemble a human, much like Japanese engineers have created with the robot called Alter, which has an embedded neural network that enables it to move by itself.[24]

The other approach to transforming the brain is the gradual replacement method, which gradually replaces every neuron in the brain with a nonbiological but perfect replacement. Supposedly this method will work, since humans naturally experience a gradual replacement process, as most of the cells in our body are continually and rapidly replaced: in a few seconds, 100 million cells are replaced. As a result, through this gradual process, each person is gradually replaced in just a few months. The premise behind this mind transfer method is that a nonbiological system can be introduced into our bodies and brains instead of ordinary biological cells, much like replacing old failing parts in a car with newer and snazzier components. In time this new brain will exist in the computer cloud. As Lee notes, citing *Slate* magazine:

> The gradual introduction of non-biological systems into our bodies and brains will be just another examples of the continual turnover of parts that comprise us. It will not alter the continuity of our identity any more than the natural replacement of our biological cells do. . . .
>
> And in the coming years, we will continue on the path of the graduate replacement and augmentation scenario until ultimately most of our thinking will be in the cloud.[25]

For this process to work, a computer has to be extremely powerful so that it can process information at the same speed as a real human brain. This is possible, since the human brain functions due to a series of electric impulses. Eventually faster computers will do this, making everlasting life possible. Since the data is immaterial, even if the physical drive that holds this copied mind were to deteriorate, the data could be moved onto another drive. If this data is immortal, so is your consciousness that makes you, you.

However, if the brain can be copied, this method raises the question of whether a complete duplicate means that more than one "you" can exist at the same time or whether the original in the aging body can be deleted like a computer file once the new version is created. Is there a way to transfer the original into a new robotic or computer form without making any copies? Might any of the data that makes up the individual be lost in making a transfer, thereby transforming the consciousness and personality? Just as files can get corrupted when they are accessed or transferred, could this happen to a copied or transferred human brain, and can this be prevented? If changes in the data do occur for whatever reason, at what point is the original individual transformed into something else?

Even though this copying approach raises all kinds of questions, from scientific to ethical issues, the ability to transform a human brain into "a digital brain with a conscious mind and all previous human characteristics and attached to a robot's body" makes immortality possible. Researchers are conducting studies on transferring animal brains in this way, and some of these brain functions have already been partly simulated, such as in the brain of a rodent.[26]

Creating Robotic or Holographic Avatars

Still another approach is creating robotic or holographic avatars as promulgated by the 2045 Initiative launched in 2011 by Dmitry Itskov, a leader in developing the Russian Internet, who has a goal of living to 10,000 years. The reason for the "2045" name is that Itskov believes immortality can be achieved by 2045, and to that end, he has brought together experts in robotics, neural interfaces, and artificial organ creation to build robotic or holographic avatars to replace the human bodies, which are subject to physical decline. As Itskov explains, in the hope of obtaining money from the wealthy to advance his immortality project, his goal is to "create technologies enabling the transfer of an individual's personality to a more advanced non-biological carrier"[27] and thereby to extend life, even to achieving immortality.

His plan involves four steps:

1. A robot or "avatar" will be developed that is linked to who we are now with a computer chip, to be achieved by 2020.
2. A human brain will be transplanted into the avatar, with a goal of doing this by 2025.
3. Human consciousness will be transferred or downloaded into an artificial brain in the avatar, replacing the biological brain.
4. Consciousness will finally evolve into a kind of global consciousness, perhaps, on the Internet, or on its own Internet, so one will no longer need a

physical presence. Instead robots, which have developed their own artificial intelligence, can now carry out any necessary physical tasks. So humanity will be evolved into a kind of brain directing these physical activities[28]

Is this possible? Is this goal what we want, even if it can be attained? Some of these possibilities do seem entirely likely, since there are already some robotic avatars that can be controlled from afar, like controlling a drone or toy helicopter. But as these robots are further developed, they might contribute to helping humans live longer by taking over jobs that entail risk to human life and health, such as working as a police officer, fireman, or miner. As they prove effective in handling the first tasks given to them, more and more avatars will be used; humans may benefit from these improvements in technology so that they no longer have to do many of these less desirable jobs. This shift in who does the actual work doesn't mean that these robots will have human consciousness, but perhaps that will become possible as researchers find way to create a copy of the human brain and transfer it into one of these avatars.

Simulating the Brain and Transferring It to a More Enduring Form

Could it really be possible to create and export the human brain into another form, including the robotic avatar? Some researchers are actually working on this, and Silicon Valley billionaires are funding much of the immortality research, as Betsy Isaacson writes in "Silicon Valley Is Trying to Make Humans Immortal—and Finding Some Success."[29] For example, Intel has been working on developing an "exascale" computer by 2018 that can operate at the same speed as the human brain. In August 2013 Japanese and German researchers used Japan's K supercomputer to simulate 1 percent of the brain's activity for one second. So just think if the next generation of computers can simulate an even greater percent of the brain's activity for a longer and longer period of time. Such research holds much promise as the beginning of expanding the ability to simulate the whole brain at the level of the individual nerve cell and its synapses with the exascale computer. Since this 1 percent model shows it is possible to do this simulation, the next step is expanding the process to encompass more and more of the brain's activity.

Spiritual Claims for Immortality

These next approaches, sometimes called "metempsychosis," are more spiritually based than scientific, but they also offer some hope of living forever for those who believe. The basic premise here is that living a better, more moral life will contribute to one having an opportunity to advance spiritually and eventually achieve immortality in this higher state of being. The belief is that the soul can jump from body to body through reincarnation, so it will continue to exist in this way. This is the belief in Hinduism, based on the idea that people live a series of lives, each one influenced by the actions in the previous one. If you live well, your next life will be better based on the law of karma, which is a spiritual law of cause and effect.

Summing Up

As the preceding discussion has shown, the science of living forever is still not ready for prime time, although the potential is there for 15, 20, 30 years in the future as multiple strands of research bear fruit.

A first step is to live a healthy lifestyle and eat a healthy diet so that you live long enough to take advantage of these new methodologies, which are in various stages of development and testing. To help you do that, a growing number of companies are promoting nutrition and fitness programs that are designed to promote longevity. Often these companies use the term "antiaging" in describing their practices or benefits, and they back up their claims with research by doctors and scientists who are part of the program. In some cases, it may seem like these programs are essentially relabeling wellness, nutrition, and fitness programs for healthy living, given the new interest in antiaging and longevity. Still, living a healthy lifestyle with a good diet will certain contribute to a longer life whether or not any doctors or scientists back a particular program.

An alternative available to anyone in the here and now, who can afford the steep cost, is cryonics—freezing one's whole body or head in the hopes of being revived in the future when the technology has developed enough to bring a person back safely. Some researchers have worked on creating an improved freezing and unfreezing process to prevent cellular and neural damage in either stage of the process. After a person has been unfrozen, other scientific procedures are necessary for rejuvenating the cells, repairing any past damages, and restoring the bodily functions, along with using any of the other biotechnology or high-tech solutions.

And then there are the spiritual beliefs about the human consciousness or soul living forever.

But the bottom line is that the realistic hope of gaining immortality lies in future scientific and research development. Today researchers are combining several different strategies as they test out different possibilities.

As described, the biotechnology paths include cryonics, curing diseases of old age, fixing the telomeres, rejuvenating the cells, using nanobots to repair the body, studying the genes for things to change or repair, using blood transfusions to provide younger blood, and creating commercial life extending products that have developed from these different types of research. The high-tech solutions include using cybernetics, creating new organs by growing them in the lab or printing them with 3-D technology, using cold storage and resuscitation techniques, downloading and uploading the brain, creating robotic or holographic avatars, and copying the brain and transferring it to a more enduring form.

Will these techniques work, and which ones show the most promise? If they work, what are the implications for society for having individuals who live much longer—or immortal lives or return from a presumed death to live again? These developments pose questions about the potential for overpopulation and limited resources as well as legal questions about dealing with estates, inheritances, and property ownership. The potential for immortality can also impact the criminal justice system: there may no longer be any murders if crime victims can be revived from the dead or prevented from dying. Family structures may be changed as well, as adults increasingly decide to not have as many or any children, because their children are living longer and because there is no reason to have children to pass on a legacy if they aren't going to die. Then, too, perhaps issues of race and class may come to the fore in deciding who is able to take advantage of these new technologies, which are likely to be expensive at first. If the opportunity for living forever becomes real, there could be a wealthy elite class of those in this category. Accidents that result in death could affect the equation as well as affect insurance payouts, which are often based on an expected lifetime of earnings. But what if that lifetime is potentially unlimited? Some scholars are already thinking about what a society would be like if people become immortal or at least live a very long time.

In the following chapters, I'll discuss in more detail some of the more developed biotechnology and high-tech research approaches and their potential impact on society.

CHAPTER THREE

The Beginnings of the Modern Search for Immortality

The modern search for immortality begins in the 1960s. By the early 1960s, there were some separate strands of research in the areas of biology, biotech, and technology that would later provide a potential path to immortality, or at least living for a much longer time. But mostly this research was confined to the lab and to scholarly exchanges between scientists working in different areas of study.

The real popular breakthrough came with the emergence of cryonics, which built on some early research discoveries to suggest these might offer hope for the future once they were developed further. Cryonics offered people something that they could do now—they could use cryogenic freezing to preserve the body through certain procedures, such as removing the blood and perfusing it with a gel to prevent the formation of ice crystals.

Thus any history of the modern search for immortality has to begin with an account of the cryonics movement and how it developed.

What Is Cryonics?

The cryonics approach is based on the premise that it is possible to live forever by freezing the body so that it can be revived in the future, when technology has advanced to the point where this revival can be successful. The cryonics procedure takes place almost immediately upon a person being certified dead, although there are some distinctions about when a person is clinically dead (the heart and breathing have stopped) or biologically dead (other organs and biological functions are shutting down). But however the point of death is defined, the key is that a person has been formally and legally judged to be dead, although preparations for that person's freezing can begin before then. Most importantly, a person who wants to be frozen has to have a will in place to permit freezing, and he or she or a designated person has to make arrangements so that a medical team will be available to begin the freezing process within minutes or at least within an hour or two of death. This quick action is imperative since initiating the freezing procedures very quickly is critical to success. Otherwise organ and cellular damage can be too great.

Based on this hope for a future revival, cryonics advocates consider the person's life before death their "first" life, since the person is expected to return to another life in the future. Another key distinction is between cryonics freezing upon death and "suspended animation," whereby a person might opt to be placed in a very cold environment or undergo freezing before being certified to be dead. From the beginning, cryonics groups and companies have approached using cryonics procedures after a person has been certified dead, so freezing the person cannot be considered murder or assisting in a suicide.

Despite cryonics' somewhat dubious and controversial beginnings, the potential for this approach actually working is much closer today than when it was first proposed in the early 1960s because of advances in medical and technology. But the early principles and concerns about the effects of cryonics resuscitation on society remain.

An Early Encounter with Cryonics

To fully understand the cryonics approach to immortality today, it is important to go back to its beginnings. Ironically I was a witness to these early days when it was first born.

This modern-day movement can trace its roots to *The Prospect of Immortality,* written by Robert C. W. Ettinger and published by Doubleday in 1964. The book created an immediate sensation with worldwide publicity in the major media at the time, which included newspaper articles, TV interviews, magazine pieces, and a national publicity tour with book signings. In response, several cryonics groups were formed, including the Cryonics Society of New York.

Around this time, I was briefly living in Manhattan from 1967 to 1968 before moving to California, and I went to one of these meetings. Though my memory is a bit foggy after all these years, I recall that it was held in the meeting room of a bank or library, and about two dozen of us sat on a few rows of folding chairs. The audience was mostly younger adults in their 20s to early 40s (I was about 24 at the time), and we were all eager to hear about the potential for immortality. The speaker, who was possibly Saul Kent, one of the movers and shakers in New York Society, laid out the vision for cryonics, emphasizing the potential for people to be frozen successfully and later restored to life, once new findings in medical technology permitted this.

After the meeting ended, I spoke with a young couple sitting in the front row, a young man and his girlfriend in their late 20s or early 30s. The young man was especially enthusiastic, and he suggested that this was a good option for someone reaching middle age. He liked the idea that one could stop aging at that point and be revived in the future, when one could not only return to life but also take advantage of the new technologies to rejuvenate the cells so that one could come back even younger and in the prime of life.

Soon after that meeting, I moved to California, so I left the world of cryonics behind, although 30 years later, an article triggered my interest, and I visited a cryonics facility based in San Leandro called TransTime. The technology was far more advanced now, and the owner showed me a large room that looked like a hospital operating room, with a long table and shelves filled with assorted canisters, bottles, and medical equipment. Then the owner pointed out one corner occupied by some gleaming silver-colored tanks of different sizes. As the owner explained, the larger tank was for a full-bodied suspension, and the smaller ones were for just preserving the head, which was a new development, since one could now be revived and reattached later to a new physical body. Plus this was a much less expensive operation for someone who couldn't afford a full-body freezing, which cost about $100,000, plus $10,000 per year for maintenance, whereas freezing only the head substantially reduced the cost—say, to about $30,000 up front and $5,000 a year for upkeep.

Then the owner explained the process they used to reduce the temperature, remove the blood from the body, and inject a gel substance to prevent crystallization, the main source of damage to the cells during freezing.

That meeting later became the inspiration for a script I wrote, then called *Bringing Grandma Back*, now called *Dead No More*, about a wealthy woman in her late 40s who is "killed" by her husband, after which he and her children take her property, while her body is placed in a cryonics tank as specified in her will and shipped off to a warehouse. However, when a young lawyer finds the woman's cryonics tank in the backyard of a house he has just bought in a foreclosure sale (later changed to a warehouse purchased from a foreclosure on a bankrupt cryonics company), he decides to bring her back with the help of a scientist friend. After they succeed and he tries to get back her property, her husband discovers she is back and tries to kill her again.

The Early History of Cryonics

To understand the hope for cryonic suspension, popularly known as "freezing" the body, as well as the fervent community of supporters that has grown up around this technique, it is important to look back at its history. Aside from the growing popularity of the programs, services, and products devoted to a healthier life to promote longevity and antiaging, the other approaches to living longer or forever are largely expressed through medical, scientific, and technological breakthroughs that are occasionally reported in the news but are largely presented as papers in scientific and medical journals. But cryonics is different. From the beginning it has built up a community of believers, almost like a cult, since the decision to have oneself frozen and the arrangements to freeze a family member lead to a continued involvement by the family and close friends who may be involved or affected should the person later be revived.

As it has developed, the early history of cryonics reads like a mystery thriller and legal drama, given the struggles of the early supporters to keep alive the belief in immortality. They have faced repeated hardships, problems in having enough money, failures in technology, lawsuits, and financial disputes. Yet most have struggled on; then new supporters have joined the movement. One of the early stories of difficulties and disasters is told in *Freezing People Is (Not) Easy: My Adventures in Cryonics*[1] by Bob Nelson, who arranged the first cryonics suspension, is being adapted into a major motion picture with a star-studded cast.

But before Bob there was *The Prospect of Immortality* by Robert C. W. Ettinger, who started the whole movement with his book describing the potential for living forever, even before the first person was frozen. So any history should begin here, though there was some early research that led to this groundbreaking book.

The Beginnings of Cryonics

When Ettinger wrote *The Prospect of Immortality,* he was working as a college teacher of mathematics and physics, so he didn't have the scientific credentials that most researchers have in working with various medical and other technologies. This contributes to the potential of living longer or forever. But perhaps Ettinger's common-man roots helped broaden the appeal of cryonics to a mass public.

He was born in 1918 to Russian Jewish immigrants. Even though he was raised in a Jewish family, he later became an atheist, so he was not restricted by the religious teachings that were skeptical of the implications of freezing the body to permit immortal life.

Ironically, his first interest in the possibility of using freezing to obtain immortality came from his interest in reading sci-fi stories in *Amazing Stories* as a teen. Later, in his early 20s, Ettinger served in the U.S. Army during World War II, where he was wounded and spent several years recovering in a hospital in Michigan. While there, his interest in the potential of immortality was peaked when he discovered research in cryogenics by French biologist Jean Rostand, and in March 1948, he wrote a short story, published in *Startling Stories*, called the "Penultimate Trump," showing how human preservation through freezing could lead to a more developed future using medical technology.

Then, in 1962, while teaching physics and mathematics in college, after earning two master's degrees in these subjects from Wayne State University, Ettinger privately published an initial version of *The Prospect of Immortality,* in which he asserted that future technological advances could bring people back to life. The book created enough of a stir that a major publisher, Doubleday, became interested and sent a review copy to Isaac Asimov, the well-known sci-fi author, who said that the science supporting cryonics was solid. So Doubleday approved the book's publication, and it became an immediate sensation worldwide. It became a Book of the Month Club selection, was published in nine languages, and was featured in the major newspapers and magazines of the day, including the *New York Times, Time,* and *Newsweek.* Ettinger soon became a celebrity and did the rounds of popular talk shows of the day, including Johnny Carson and Steve Allen.

The result was that numerous groups sprung up around the country, which formed the beginning of the cryonics movement. One was the Cryonics Society of New York, where I first learned about cryonics in 1967. Many of these societies also owe their founding to the work of the lesser-known Evan Cooper, who wrote a book called *Immortality: Scientifically, Physically, Now* in 1962 under the pseudonym of Nathan Duhring. It made the same claim for cryopreservation as Ettinger made, although it didn't have the same scientific and technical foundation and was never published by a major publisher. Still, Cooper helped generate interest in freezing by becoming the first activist for cryonics, even though that name was not used until 1965, and his efforts led to forming the first cryonics organization called the Life Extension Society (LES). The organization urged immediate action to start preserving humans through freezing, and he set up a national network of chapters with coordinators who could begin the process.

Whereas Cooper, who dropped out of the movement in 1969, initiated these societies, (and the meeting I attended may have been one of these), Ettinger's book soon became a driving force in sparking nationwide interest in cryopreservation and participation in these groups. For example, in 1966, cryonics societies were formed in California and Michigan, and Ettinger became the president of the Cryonics Society of Michigan. Even so, these early years from the 1960s through the 1980s were fairly slow and somewhat tumultuous for the cryonics movement, largely because of a lack of solid support from the medical, business, or financial communities. As a result, most of the efforts to find candidates for cryopreservation were slow, and some of the first efforts were subjected to mishaps when the technology failed, leaving a half-dozen individuals in suspension dead. Yet somehow the movement struggled along.

What did Ettinger say in his book that was so compelling? And what happened to slow down progress in these early years? The next sections describe these developments.

The Power of Ettinger's Book

Ettinger's book was so influential because he brought together all of the research that suggested that the methodology could now be applied to freezing humans, along with considering the implications of this new technology for society, religion, the law, and economics. Ettinger also considered the issue of identity and what a freeze-centered society might look like. The book came along in the mid-1960s, at a time when society was undergoing a major transformation that marked the beginning of the "me" generation, when the first baby boomers were turning 20. So it was a time when society was ready to change, reflected in the protests over the war in Vietnam, the women's liberation movement, and the civil rights movement. It was a time when all kinds of new things became possible, and the potential for immortality might have seemed like one more type of transformation. When I went to the cryonics meeting in 1967, the feeling of breaking free of the past to find something new was in the air.

The Case for Cryonics

Importantly, about the first third of Ettinger's book is devoted to making a case for the biology of making freezing and reviving humans possible, and he writes in simple, direct language, creating a compelling argument for the process that the average person can understand. He emphasizes at the beginning of the book that the potential for future immortality exists right now if we only store ourselves in suitable freezers, so we can be brought back when medical science has advanced such that it can repair any damage and cure the body of the ills that cause death. He points out that "at very low temperatures it is possible, *right now*, to preserve dead people with essentially no deterioration, indefinitely." He suggests that in the future medical science should eventually be "able to repair almost any damage to the human body, including freezing damage and senile debility or other cause of death." Thus it is only necessary for people to arrange to have their bodies frozen in suitable freezers after death until science can revive and cure them.[2]

However, it has to be applied very soon after the person is biologically dead so that the condition of most cells might closely resemble those of life. Otherwise, if cellular death has occurred, the individual cells in the body have experienced irreversible degeneration so that a suspended death through freezing is no longer possible. However, later research has suggested that the brain may survive the decline of the body and so could be revived with a different or largely mechanical or computer-based body.

Ettinger then highlights the various strands of research that show that suspended death is now possible. For example, he describes some successful small-scale freezing techniques that have been conducted at a number of laboratories and hospitals in the United States, France, Britain, Russia, and other countries. He compares suspended death to being under anesthesia in the hospital for a very long time, even for centuries.

Then Ettinger provides the case for long-term storage at freezing temperatures, describing the existing technology of the time, to show that a body cooled by liquid nitrogen might be stored without significant changes or deterioration for years or even centuries, while a body cooled by liquid helium might be kept forever. He next provides supporting evidence from animal and human research to show why this storage arrangement works, such as Professor Jean Rostand's research showing that the movement of frog spermatozoa was preserved for several days at $-4°C$ to $-6°C$. In another experiment, Andjus and Loveland reported that 80–100 percent of ice-cold rats recovered and survived.[3]

At the same time Ettinger points out the potential sources of freezing injury, which can help future scientists seeking to find fixes for these problems, resulting in a successful freezing. Among these are

- mechanical damage by the ice crystals, such as expanding and breaking the cell membranes and bodies, although a rare event
- a dangerous concentration of electrolytes during the freezing process when the fluid still in a cell has an unnaturally high concentration of salts and similar substances called "electrolytes"
- a metabolic imbalance due to the unequal effect of cold in different parts of the body.
- thermal shock causing some cells to die, since various materials in the cells and their membranes shrink at different rates as the temperature is lowered
- damage during storage, since eventually some changes occur, even at the very low temperatures, although they may be very slow
- damage during thawing, the main source of damage, especially if the thawing is slow and protective infusions aren't applied
- other deleterious effects such as drugs, antibiotics, and normal body solutions that become concentrated at lethal levels or incomplete freezing if glycerol is used at a dry ice temperature

Ettinger similarly describes at length the early research on protective infusions that prevent or reduce freezing damage, concluding that the most satisfactory substances were glycerol and dimethylsulfoxide. Thu future technicians would only have to develop the necessary methods to successfully thaw and remove the protective agents, since there would be little freezing damage and the technicians would not have to engage in exceptional efforts to reverse freezing damage.[4]

To respond to concerns that the brain might be wiped clean of memories, even though the body was restored to an active life, Ettinger also shows that one's memories should persist after freezing, drawing on the research on mammals and humans to show that this should not be a concern. For example, Dr. A. U. Smith reported in a 1961 book *Biological Effects of Freezing and Supercooling* that rats trained to find food in mazes showed no appreciable memory loss after they were cooled to a body temperature just above freezing.[5]

Thus even 50 years ago there was evidence that the damage from freezing could be overcome, with the prospect of even further methods of overcoming this damage to be developed in the future. For example, he suggests that it will become possible to grow or enable the body to repair itself by regenerating missing parts, and he believes that only a small percentage of brain cells need to survive with little damage so that scientists will have enough cells to faithfully reconstruct the brain with freshly generated tissue. Accordingly, as soon as possible after death, and even within an hour or two, the process of preparing and freezing the body should begin, as long as any cells in the body show life—a period measured in at least hours and sometimes days.[6]

He points to a number of techniques that were already being used to keep freshly dead bodies in good condition in order to maintain the organs in good health, in case a transplant couldn't be performed immediately. For example, in 1963, heart-lung machines were used to keep the body supplied with oxygenated blood for up to 18 hours after death, before livers were taken from the bodies for a transplant operation. Another technique was putting a patient in an operation inside a pressure chamber, to breathe oxygen so that surgeons could stop the blood circulation for twice as long as normal.

He also discusses the properties of different cooling agents for lowering the temperature of the body. For example, he points out that a dry ice temperature of –79°C can slow down or halt all biological processes, and he concludes that initially the best substance to use is liquid nitrogen; once permanent installations are built, liquid helium might be used instead, although in an emergency or for a lower cost one could use dry ice, which is cheap and easy to handle.[7]

The Potential for Revival, Mechanical Aids, Transplants, and Regeneration

After making the case for freezing the body, Ettinger discusses the potential for reviving it in the future. Among other things, he points out how hundreds of people have been revived after being clinically dead for many minutes (heartbeat and breathing have stopped) by using fairly simple measures, including artificial respiration, blood transfusions, heart massage, and stimulation by drugs or electricity. He suggests that in the future, some repair can help stave off death due to the failure of some vital organ due to disease or old age by repairing or replacing the defective organ that caused death, curing any acute disease or making other needed repairs, and finally making a general overhaul and rejuvenation.

He also points to inventions that perform biological functions that were available in the early 1960s, such as respirators, oxygen masks, pressure changes, and iron lungs to aid breathing and electronic pacemakers to help a heart keep proper time. Plus there were machines that could do the work of missing or diseased kidneys as well as all kinds of prosthetics.

He even cites a professor of biochemical engineering, Dr. Lee B. Lusted at the University of Rochester, who predicted that in 50 years it would be possible to replace nearly all of the organs in the body (such as the heart, kidneys, stomach, and liver) with artificial organs or with built-in electronic control systems, and of course advanced artificial arms and legs would be available. At the time there were already some beginnings of these developments, such as the first human lung transplant in 1963 and a number of successful kidney transplants[8]; today the field of organ transplants has progressed tremendously so that even hearts and whole face transplants have become possible.

Besides pointing to the beginnings of human transplants, Ettinger notes that organ culture and regeneration could be used to take a germ cell or ordinary somatic cell from the body of the resuscitated individual to grow the organ or organs needed for repair. He also considers the potential for curing old age and repairing the brain, again sounding very much ahead of his time. For example, he proposes that the brain could be rejuvenated by brain cells grown in the lab and then given the appropriate information, after which surgeons could use them to replace senescent cells. Over time these cells could be periodically replaced.[9]

Additionally, Ettinger considers the various strategies for living forever, such as preserving samples of ourselves by growing cultures from frozen cells and growing future organs, which will provide healthy organs when needed for aging individuals to keep them young. Ettinger even suggests that this approach of preserving cells might be used to take tiny samples from many regions of the brain (noting their location as accurately as possible) so that scientists might later duplicate the memory traces in various regions of the brain as well as learn to read the messages encoded in the brain's network of nervous tissues. The premise is that scientists in the future will be able to map the brain and its contents and fill in any gaps in a person's memories with other information.[10]

The Steps to Take Now

Having made his case for how the biology and biochemistry of living forever through freezing work, Ettinger provides a guide for what individuals should do now to make sure they are frozen. No wonder his book is considered the bible of the cryonics movement.

First he recommends that individuals specify in their will, drawn up with competent legal counsel, that they want to be frozen, and he notes that a number of people have already done so in several states—Michigan; Washington, D.C.; New York; New Jersey; and California—and in Japan.

They should get a promise of cooperation from their expected surviving next of kin (preferably in writing), choose a supportive executive, provide funds to carry out their wishes (possibly from a special insurance policy), and obtain a promise of cooperation from a physician to take the needed steps to help the person get frozen upon death. Ettinger also foreshadows the rising cryonics movement in suggesting that many organizations will sprout up in the fairly near future, offering various services in connection with the freezer program.[11]

The Implications of Cryonics for Religion, Law, and Society

The book also includes a chapter on how freezing the body can be reconciled with the principles of major religious groups. For example, Ettinger points out that clergy of most denominations now hold to the view that the advance of science is compatible with a belief in God, and he believes that clergy will eventually come to see freezing as a way to save or prolong life. Further, he seeks to reassure the religious faithful that preserving the body through freezing represents no danger to religious institutions. Rather it can be viewed as part of God's plan and that all religions continually reexamine and adapt their faith in light of new discoveries and new capabilities.

Likewise Ettinger seeks to show how freezing and reanimating the body can be compatible with the law. Most naturally this approach might be included in the system of laws governing the disposal of bodies and the operations of cemeteries, mausoleums, and home crypts. He points out how present laws generally give priority to the wishes of the deceased and next of kin and that there is legal precedent to allow for the unusual treatment of bodies.[12] A modern-day parallel is the case of Bethany Maynard, who moved to Oregon to use the assisted suicide law there to obtain medical assistance to commit suicide because of her terminal illness, before California law was changed to permit this in response to the news of Bethany's case.

Presuming freezing is allowed, Ettinger then discusses the legal questions that might occur about the rights or obligations of the corpse. For instance, these rights might include protection of his body and property, governmental supervision of the freezer and of his trust funds. In fact, Ettinger suggests the law may need to distinguish between three classes of people—those in suspended animation (such as those in a coma), those frozen after death, and those who are really dead because there is no way to bring them back. Then there is the question about what happens to supposed widows and widowers if the ^rescuscitees come back to life, and the surviving partner now faces a younger revived partner and an older still-living mate. Maybe multiple marriages will need to become a possibility to preserve the rights of the once-dead along with the living.[13]

In short, Ettinger recognizes the potential that freezing has to change the structure of society, and he discusses the need to change the laws to accommodate these changes, since the frozen will be potentially alive and therefore will be property owners and taxpayers.[14]

What of the burgeoning population of the revived dead? In Ettinger's view, there will be a glorious future in which the globe will have found ways to increase the resources available to society. Machines will be able to do much more, even reproduce, duplicate themselves, and improve themselves, much like computers have been used to help in designing new computers. It is a society that will rapidly grow richer (if based on the continuation of the progress already achieved in the past century). By the same token, he claims that extended longevity won't contribute to the world population problem that already exists due to natural increase. Instead he asserts that the freezer program will contribute to creating a reasonable birth control program, and even a general eugenic program, since knowing of the potential for immortality will tend to make people more responsible in all areas, including population control. Moreover, should crowding become an issue, he proposes that people could agree to share the available space in shifts so that the same people might go into suspended animation from time to time to make room for others.[15]

So what about the costs of freezing? Ettinger treats this like a question of supply and demand, suggesting that as more and more and more bodies are frozen, the costs could be reduced by storing a sizable number of bodies together, such as in a specially designed or modified building. As for paying for these costs, he suggests that before being frozen, people will seek to safeguard their bodies while frozen and ensure that they will be reintegrated into society on their return, such as by setting up a special trust fund to cover the preservation costs. In his view, this potential for freezing will go global: the possibility of freezing will expand across borders around the world, regardless of political ideology. At the same time, less expensive methods of freezing will enable the procedure to be used in poorer countries, and worldwide cooperation will help share resources.[16]

A Question of Personal Identity

Ettinger additionally addresses one of the philosophical and ethical issues raised by the promise of immortality—the issue of identity. This is a core issue because the resuscitee may emerge from being frozen with a patchwork of new organs and parts. He might have internal organs, such as his heart, lungs, liver, and kidneys, which have been grown from someone else's donor cells. He might have arms and legs constructed from fabric, metal, and plastic that operate by tiny motors. His brain cells may be mostly new, regenerated from a few cells that could be saved, and some of his memory and personality traits may have been imprinted onto his new cells by methods from chemistry or physics.

Ettinger concludes by suggesting that individual identity is only partly existent and partly invented. Rather than having an identity, we have degrees of identity. Given all these modifications, one's very identity, composed of our memories, personality, and abilities to think and feel, might be changed in the process. Plus this changed individual has to adapt to a brand-new world. Thus, we might not come back as our old selves, but as someone new.[17]

Toward a Glorious Future

Ettinger even suggests that those who are resuscitated may possibly come back as superior individuals, since somatic improvements may enable them to live much better lives. It might be possible collectively or individually to design ourselves by selecting the desired traits and abilities. Ettinger further suggests that most people living in the future world will want to be frozen by nondamaging methods and that the reversal of aging will be easier than completely designing and rebuilding the body. As a result, this ability to freeze and revive people in a time of new technologies will lead to lead to a glorious new future for humanity, characterized as a time of peace, even though 50 years later these future hopes seem hopelessly naïve given the tumultuous times of today. Among other things, Ettinger imagines this glorious future will have these characteristics:

- The prospect of immortality will help reduce impetuous actions and antisocial behavior, since national leaders will want to preserve themselves and their societies and so take a much longer view of potential conflict.
- People will behave better with others, because they will realize that not only they but others will be around for a long time, with paths that cross repeatedly.
- The medical arts will be so far advanced that few types of accidents can result in permanent death; rather, people will have a cure or replacement options available to fix most illnesses and injuries.[18]

In short, Ettinger imagines a future golden age that will bring unlimited wealth, since matter and energy will be freely available, while thinking machines will be virtually unlimited in what they can do. Additionally, all kinds of new goods, services, and modes of living will be available, cities may even be weather controlled, and automation will either eliminate dull and unpleasant jobs or workers will have shorter hours or higher pay.[19]

In short, Ettinger envisions a future utopia made possible by freezing. He believes such a program is inevitable for several reasons. First, as soon as suspended animation is practical, people with incurable diseases will want to be frozen alive to await the time when cures are discovered, and new technologies and growing wealth will make it very tempting to want to do this. Second, most people will see the value in being frozen with the hope of returning successfully and being educated and adaptable to live in this future world. Third, a freezer program will resolve all kinds of problems of today, from disease prevention and human relations to international relationship, because it provides a longer time period for reaching a solution. It can even help end nuclear war, because the reckless usually have little to lose, but now everyone will have a great hope for the future after being revived from the freezer.

Thus his book is like a clarion call for those who believe in the potential of freezing humans, based on his explanation of how this can be done and why it should be, to work toward the reality of this happening. For an individual this means choosing to be frozen at the time of death—or even before through suspended animation—to await being revived when the technology is developed to experience a new life in a new world with tremendous potential. For society as a whole, this can mean working toward the acceptance of this practice; more and more people will want to be frozen and will support those who choose to do so.

Reading his book over 50 years later, his blueprint for a new society based on successfully freezing people and eventually bringing them back to a renewed life in an age of prosperity seems hopelessly unrealistic. Yet it is important to see this book as the fundamental work in building the hope for cryonics among a growing number of adherents. In turn, this belief in the cryonics future has contributed to the continued advocacy for cryonics that has led to the rise of different centers and societies devoted to the hope that freezing will ultimately lead to immortality. Today about 250 individuals are now frozen; many more are part of a community of supporters, among them those who plan to be frozen in the future.

Yet despite this glowing hope for a golden age offered by the potential of cryonics, its early history is anything but. Rather it is like a mystery crime thriller, in which different factions of believers struggle with the law, society, and each other in order to realize the cryonics vision while confronted by the limitations of the technology of the day. Unfortunately these limitations have resulted in the death of some of the first frozen individuals, some lawsuits by family members and business associates over what went wrong, and a continued popular skepticism about the viability of this approach, particularly since no one has yet been revived—there are just these heads and bodies of believers before they died, awaiting their revival in the future. I'll next discuss what happened in these early years, since these battles also helped prepare the way for the modern science of cryonics.

Freezing the First Cryonics Patients

One of the most dramatic early stories about the struggles with cryonics was what happened to Bob Nelson, a TV repairman, who was inspired by *The Prospect of Immortality*. He joined one of the early life extension groups, and in 1967 he worked with a biophysicist to freeze the first human to be placed in cryonics suspension—Dr. James Bedford. While Bedford is still in suspension, waiting to be revived, other individuals frozen by Nelson were not so fortunate due to a cyber malfunction, leading to a big lawsuit against Nelson's company. Even though Nelson had contributed his own savings to preserve the patients, that wasn't enough, and he ultimately lost the suit, while efforts to preserve other patients went on.

His book is a fascinating story, which shows the difficulties faced by the first cryonics companies and their patients hoping for a future life. It also has parallels with what happened after the Wright brothers launched their first flight in a rickety plane in Kitty Hawk, North Carolina, in 1903. Their first three-second flight attempt on December 14, 1903, resulted in the flyer stalling after takeoff. Their next three flights on December 17, 1903, witnessed by five people, were only for 120, 175, and 200 feet at less than seven miles per hour. The flight gained an altitude of only about 10 feet above the ground and lasted less than a minute. In 1905, their next flying machine had problems as well; its first flights on June 23 lasted no longer than 10 seconds, and on July 14 Orville suffered a serious crash.

After a few months in September and October, they did slightly better after adding more improvements for the rudder and wings, which made the plane more stable, so they had longer, more successful flights lasting 17–38 minutes. Yet they had little press coverage and much skepticism, in part because no reporters or photographers witnessed their first flights and because they were fearful of competitors stealing their ideas. So they continued to work largely in obscurity, while other aviation pioneers began working on their own designs.

The press and public were highly skeptical, questioning whether they had flown as they claimed, as they continued to slog on in trying to interest buyers from the U.S. military to the kings of Great Brain, Spain, and Italy. Finally the tide turned in 1908, after they improved their technology so they were able to fly for an hour on September 9, and they began taking passengers up for rides. The rest, of course, is history.

The early days of cryonics—and Bob Nelson's role in these first freezings is a little like that—a struggle to get the technology to work and then to gain interest from prospective clients and the media. Unfortunately, he met with failure after he became one of cryonics' most active early spokesmen and the world turned against him, due to some highly publicized failures.

How Bob Nelson Got Started

Bob Nelson first got involved with cryonics in 1965 when he was living in Los Angeles, after separating from his wife, and working as a TV repairman. While swimming at a pool at an apartment building, he saw a newspaper with an article about a Michigan professor who had written a book called *The Prospect of Immortality*. Later he reread the article again and again and kept thinking about how cryonics could change life and death and how this technology could enable him to achieve a childhood dream of traveling to other planets and seeing civilizations of the future. He thought about how suspended animation could allow people to stay dormant for months or years as they traveled to faraway solar systems and galaxies.[20]

The book also led him to think about death in a new way. He now viewed it as more of a process than a single event, with a series of biological steps along the way to a person being fully dead. As a result, he reasoned, lowering the body's temperature at an early stage of the dying process—at clinical death, not biological death—would stop the progression, so the patient would never reach irreversible death.

After receiving feedback from some associates, Nelson got some help from an official with Mensa (the society for those in the top 2 percent of IQ) to set up a legal organization and publish a newsletter to educate the public about this new science. He also began to think about the many examples of suspended animation in nature, such as a snake caught in a blizzard in southern Alaska that is frozen solid and covered by several feet of winter snow. But in the spring several months later, the snake reemerges and slithers away to its breeding grounds. This was another example of the way that life could be frozen for long periods of time and be revived in a more favorable environment. Now Nelson realized that humans had the chance to do the same through low-temperature biology.

For Nelson, this realization led him to want to join the cryonics movement. Then, after he heard a radio DJ mockingly announce that any listeners who didn't like the idea of dying someday could call Helen Kline for an invitation to the first meeting of the Life Extension Society, Nelson attended California's first LES meeting on May 13, 1966. There he met Helen Kline, an elderly woman who was wasting away from the last stages of lung cancer, and many other LES members. Many of the people were elderly too, but all were hopeful of the potential for immortality. At the meeting, they listened to an audiotape of Ettinger appearing on the popular late-night *Johnny Carson Show.*

Four weeks later, Nelson was elected president at the group's second meeting, and he felt that leading this group was his legacy in life. A few months later, he incorporated the nonprofit Cryonics Society of California. After that, he was visited by the Cryonics Society of New York's President Curtis Henderson and Secretary Saul Kent—the same group where I first learned about cryonics. The New York group had already existed for a year, and Nelson's California group adopted its same organization and structure. Kent and Henderson gave Nelson's group an education on the state of cryonics and advised them to find a cooperating mortuary and medical and hospital officials that would honor a patient's dying wish to be suspended. Soon afterward the group received its corporate approval and nonprofit status and set up offices in Westwood, California. Over the next six months Nelson and other group members appeared on several radio and TV programs promoting cryonics. The group also gained the support of a prestigious scientific advisory board. Then, after the group raised enough funds, they invited Ettinger to come and speak to them about his thesis of extended life.

Meanwhile, as several other cryonics groups sprung up in response to Ettinger's book, sought new members, and worked on setting up storage arrangements for the expected flood of patients wanting to be frozen, Nelson decided to focus on research to protect patient from the damage caused by freezing. Among other things, he met with Robert W. Prehoda, a medical researcher and prolific author on cryobiology, who introduced him to meetings with other scientists, which eventually led him to form the scientific advisory council, with scientists who believed in the potential of research to achieve groundbreaking discoveries in cryonics. The plan was for the Cryonics Society of California to provide funding while Prehoda helped obtain grants for this research.[21]

The First Big Freeze

However, despite the hopes for a promising future, by December 1966, no one had been frozen. That changed about a month later, when Dr. James Bedford's son Norman sought out a mortuary, because his father wanted to be frozen and suspended. The mortuary, the Stedford Mortuary in Glendale, California, then called CSC. Though the advisory board was reluctant to remain involved if there was an actual procedure, Prehoda and Ettinger were encouraging, since someone of Bedford's position—a university psychology professor—would give the emerging science a new credibility. Even if the CSC lost its scientific board, Ettinger urged Nelson to go ahead, since freezing the first man would be a major accomplishment and would gain worldwide attention. And probably after that any advisory board members would return.

With Ettinger's support, Nelson decided to freeze the first human being, and Dr. Bedford's son assured him that he could obtain Bedford's written authorization along with $300,000 from his father's foundation for cryobiological research to cover the expenses of freezing his body. Among these expenses, were the chemicals for the procedure, a temporary dry ice storage container for the patient, and a second container to be filled with liquid nitrogen for the permanent suspension. Plus there would be a cost to store the container.

Though the specific procedures have changed due to scientific progress, the basic goals of the procedure were the same: protecting the brain (the only irreplaceable organ) and the cells of the body. The chemicals they used included glycerol with dimethyl sulfoxide (DMSO) that was injected into the bloodstream to absorb about 90 percent of the cells' water so that ice crystals would form outside the cells rather than inside.

Given Bedfod's assent, on January 12, 1967, the first human freezing went forward. After an attending physician, Dr. Renault Able, witnessed Dr. Bedford's heart stop beating and pronounced him deceased, the perfusion team injected a biological antifreeze solution with 15 percent DMSO into Bedford's body. As the cryogenic cooling process began, the perfusion process continued for two hours, after which the cryonics team lifted Bedford's body from the pile of crushed ice on the operating table and onto a much colder dry ice at –109°F inside a wooden, polystyrene-lined container. After they covered his body with more dry ice, they immersed his body in liquid nitrogen at –320°F. At this extremely low temperature, all chemical and biological activity would finally almost completely stop, and presumably any tissue would not decay for hundreds or thousands of years.[22]

So that was the beginning of freezing anyone, and the procedure immediately became big news. It also scared members of the scientific advisory council, since they were afraid to be associated with this new, and now very controversial, procedure, which headlines splashed around the world. In response to the high level of interest, Ettinger flew to California to assist with the publicity outreach, which included a press conference with the CSC and at least 30 news media and television interviews. The procedure also became the lead story in an issue of *Life* magazine, and that's the point at which I learned about the procedure and went to my first cryonics meeting. Additionally, Ettinger appeared on most of the big celebrity talk shows, including the *Johnny Carson* show, while Nelson appeared on the *Phil Donahue Show* and several others. While the audiences seemed curious, many wondered whether a greatly extended life was really possible, and some mistakenly viewed cryonics as a replacement for an afterlife rather than as a way of delaying death.

Thus, while the first freezing occurred with much fanfare, the immediate aftermath was a rather bumpy road ahead, as the cryonics society and the movement generally faced problems with funding, infighting, and some well-publicized failures. It took about 20 years, until the mid-1980s, with the emergence of the Alcor Foundation and other commercial cryonics facilities with state-of-the-art technologies, for the movement to overcome its early struggles.

The Long Struggle to Survive

After the frenzy of news reports and interviews died down, Nelson and the CSC decided to focus on seeking new patients who wanted to be frozen. It seemed to make sense to do so after a mortician, Joseph Klockgether, called, since he thought cryonics could potentially become a worldwide phenomenon by changing the definition of death. Around this time Nelson also met Marie Sweet and her husband, Russ, who were both elderly and wanted to be frozen after their deaths.

Ironically, after Marie got 20 scientists to write papers in biology, physiology, and space exploration for the First National Cryonics Conference to be held in Santa Monica in August, 1967, she died the first night from heart failure. Even though her body wasn't discovered for two days, Nelson didn't have the legal permission to do the procedure, and the CSE wasn't able to reach her husband to sign the formal authorization, Nelson gave the go-ahead, since it was now or never. The decay process had already started with bacteria already attacking Sweet's body and her blood pooling due to gravity. But rather than wait any longer, Nelson arranged for her body to be cooled down, transported her to Klockgether's mortuary, and cooled down even further to subzero temperature. In this way Marie Phelps-Sweet became the second person to be cryogenically frozen.

Fortunately, her husband, Russ, approved, feeling that it meant so much to Marie to be frozen, even though these weren't the ideal conditions for a cryonics suspension because of the two days between death and freezing Marie's body. Still Nelson was hopeful that future scientific discoveries would make it possible to reanimate her, despite the two-day delay. But fiscal realities set in when Nelson told Russ that the cost of finding a capsule, a permanent storage facility, and a liquid nitrogen provider would cost at least $10,000 to start and maybe more. It was beyond Russ Sweet's ability to pay; he only had $300, as he and Helen had mostly been living on their social security, despite living in a beautiful home overlooking the Pacific Ocean. So Nelson agreed to cover the costs, hoping to recoup this through donations from the cryonics community. Even though Nelson eventually raised some funds, he had to pay most of the money himself, and sometimes he had to choose between paying his mortgage and paying for dry ice for the next two years. Yet he did so, believing fervently in the hope of future reanimation and feeling that maintaining the first patients in suspension, even if they didn't have the money to afford it, was a necessary step toward that goal.[23]

In May 1968 Nelson had another cryonics candidate, Helen Kline, who had hosted the first California cryonics meeting. Her sister called to say that Helen was in a coma, and her condition was fast deteriorating. At least Helen had signed all of the necessary documents for her body to be frozen at her death; unfortunately she didn't have any money either. Still, Nelson gave the go-ahead, and after the doctor declared her legally dead, she was cooled down and frozen.

In September 1968, when Russ Sweet died after a heart attack, the hospital staff ignored the documentation Russ carried with instructions to cool his body with ice the moment he died and to contact the CSC immediately. Instead they contacted Russ's next of kin, so eventually six hours elapsed between Russ's time of death and the time that perfusion began at Klockgether's mortuary. Memory loss was likely if Russ was ever reanimated, even though suspension was still considered viable. However, Russ also didn't have enough money for perpetual maintenance. He only had $10,000 to cover his suspension.[24]

Unfortunately these early decisions set the stage for problems later on, since these suspensions were either imperfect, insufficiently financed, or both, and they suffered from the limitations of the technology of the time. Yet at least the patient might have a chance, though uncertain, of later being revived to live again.

Spreading the Word

In any event, by September 1968 the CSC membership had grown to nearly 100 people. Imagining that they would now get a huge number of requests for cryonic suspensions, Nelson and the group set up a for-profit corporation, Cryonic Internet, to design, develop, and provide cryogenic hardware equipment. The plan was for CSC to contract any perfusions to CI so that they could store members who were frozen. Plus now these people wanted a permanent storage facility, since Klockgether wanted the bodies out of his mortuary.

The CSC found a place to store the bodies by creating an underground vault under a cemetery, since the facility would no longer be subject to local laws affecting the disposition of frozen patients. Being in a cemetery provided some legal cover in case the experimental technology for suspensions failed. After all, if the patients were in a cemetery, they would be considered "legally interred," so even if the cryonics program ended, they could stay in the vault unless their still living family members wanted them somewhere else. Accordingly, the CSC made arrangements with the Oakwood Memorial Park Cemetery in Chatsworth, a suburb of Los Angeles, to store them.

Soon Nelson contracted with a company to build a 12 ft. by 18 ft. vault that could hold up to about 20 bodies. It was constructed with steel and cement, much like a swimming pool. Then Nelson found a capsule to hold up to 30 bodies at an industrial liquidation yard in Los Angeles, which had all the instruments needed to monitor the temperature. Even though the $5,000 cost was higher than the $2,500 remaining from Russ's donation, and it would be costly to maintain the bodies with a continuing supply of liquid nitrogen, Nelson imagined that the CSC could handle the financial arrangements with five or six more paying suspensions. So he took out a loan to pay the balance, while the cemetery owner agreed to store the capsule in its heavy-equipment yard until the CSC was ready to use it.[25]

Thus Nelson was very hopeful for the future success of cryonics, as he continued to travel around the country giving presentations to sell cryonics and their new facility to the public. But in late 1968, Nelson had more problems, when Marie Brown contacted Nelson about her father Louis Nisco, who had been in cryonic suspension for a year at the Cryo-Care Equipment Corporation in Phoenix, Arizona. She was desperate because she could no longer afford paying for the monthly storage and liquid nitrogen maintenance, and the company owner, Ed Hope, was threatening to terminate the suspension. So again Nelson came to the rescue. He offered to pay the remaining balance that she owed on the capsule and take charge of her father's suspension, in return for her donating her father and the capsule to the CSC and making the $150 a month payments to cover the costs of storage and liquid nitrogen. Unfortunately, once her father's body was removed, she never made the monthly promised payments. Instead she sent a letter that she was leaving her father's fate with Nelson and the CSC. Then, making matters worse, Marie's father's capsule began leaking badly, and it needed a vacuum pump to maintain a vacuum between the outer wall and the inner chamber. Also, the liquid nitrogen had to be replaced every week instead of once a month. Still Nelson persevered, and in March 1969, he arranged with Klockgether to open Louis Nisco's capsule and place the other three patients they had frozen—Marie, Helen, and Russ—inside. Then, he and the CSC moved the capsule to the Oakland Memorial Park Cemetery in Chatsworth and placed it in the heavy equipment yard. Still, despite the many setbacks, Nelson was determined to press on, feeling that failure was not an option.[26]

But now everything seemed to go wrong. It was as if the limits of technology and money indicated that the cryonics approach was not yet ready. Nelson described the grim situation that existed after fixing the vault over the next year:

> About a month after the vault was completed, water started seeping through the walls . . . Six months after construction, the vault was accumulating a foot of water per day . . .
>
> Time passed, and my early hopes faded. The predicted boom in cryonics patients never materialized.[27]

The irony is that Nelson had completed the vault and was ready to permanently store patients, but both he and CSC were broke. He had spent two years and several thousand dollars caring for a handful of patients, and now felt it was almost the end of his dream and the fate of the four patients sealed within the capsule as the remaining liquid nitrogen boiled off. He had no money, and he seemed to have no hope of getting any.[28]

Then in September 1970, shortly after he felt all was lost, he got a call from two brothers in Des Moines, Iowa. Their mother Mildred Harrington was dying of cancer, and they wanted to know if she could be frozen. So he flew to Iowa to meet them, and within a few days, he found a cooperating doctor and mortuary. A week after he arrived, Mildred died, and with the help of Ettinger and Klockgether, the perfusion was performed. After that, the brothers turned into media hounds for the local media. Klockgether transported Mildred to the underground vault at his mortuary, where Nelson had to replace her dry ice weekly with at least 300 pounds of new ice. Then, two years after Mildred had been in temporary storage in the vault, the brothers asked to hold a two-hour memorial service for their mother and show her off in her frozen state to about 10 relatives. So Nelson and Klockgether designed a special casket with a spray valve to keep her body at a low temperature to avoid thawing during the two-hour service. Finally, the day of the memorial service and viewing arrived, and with a black wig, false eyelashes, and embroidered white gown, Mildred looked something like a Sleeping Beauty, as 10 guests observed the ceremony, leading Nelson to feel hopeful again that cryonics would gain a wider acceptance.

Then Nelson had another patient for suspension, when Guy de la Poterie called from Montreal, Canada, in July 1971 to tell him about his seven-year-old daughter Genevieve, who had cancer and had only a short time left to live, since her remaining kidney was failing. But again, Nelson had a problem of a family member with no money, although Nelson felt it was his duty to help. So he spoke to the hospital director, who agreed to discharge Genevieve, and after some difficulty in finding a mortuary in Canada to handle Genevieve's case, Guy called to say he was in California, and Genevieve was accepted into the Children's Hospital in Hollywood. Even though Guy was unable to get the $10,000 from his family to cover the costs, Nelson once again accepted a cryonics suspension without pay, feeling he couldn't say no. [29]

Nelson even took Genevieve to Disneyland a few weeks before she died. This time, Nelson had the usual perfusion filmed by a professional cameraman, and the film was later used in a documentary. Since Genevieve didn't look dead, merely asleep, Nelson felt even more confident of cryonics' potential. Then he returned to educating the general public about cryonics, thinking of Genevieve and the other frozen patients as akin to heroes who were counting on him for their ultimate victory over death.

Meanwhile, after Nelson had frozen the first man, several other groups, including the Cryonics Society of New York, had frozen a few patients, even though they suffered many of the same problems—mainly problems in finding a place to house the bodies, leaks in the capsules so that the liquid nitrogen evaporated, and deteriorating bodies. Nelson encountered more problems due to individuals arranging for cryonic suspension for a dying parent but then not paying for their continued suspension, leaving Nelson with the problem of what to do since it was not just a financial issue, like getting a car repossessed for not making the payments. Instead, since Nelson felt he was dealing with a patient who might one day live again, he viewed suspension as a matter of life and death. As a result, he ended up trying to keep even more patients "alive," even though their relatives had ceased to pay the bills; he was only able to do so because he received a small income from cosmetic preservation and storage that helped him pay for nonpaying suspensions.[30]

Unfortunately, Nelson faced more than financial problems; the failings of technology resulted in meltdowns when the capsules no longer maintained the cold temperatures needed to keep the bodies frozen so that they began decomposing. He had already lost one patient that way, and in the 1970s he lost still more after he decided to move his last two suspensions of Genevieve de La Poterie and Mildred Harrington from dry ice and put them in a capsule with another patient. To do so, he opened up Steven Mandell's capsule and put the bodies of Genevieve and Mildred with him.[31] But even though he monitored the capsule each day, the design was flawed so that the capsule acted like a leaky faucet in which the liquid nitrogen gradually leaked out, while high 100-degree temperatures in Chatsworth during the summer interfered with the pump's optimum performance. Unfortunately, in 1974, 10 years after Nelson had taken over as the president of the CSC, when he went to check on the capsule holding Steven, Genevieve, and Mildred, he discovered the vacuum pump was no longer running, and when he touched the vent, it was not just warm but hot. This meant that the bodies inside were certainly decomposing. The problem developed after the cemetery groundskeeper discovered the pump stopped working while Nelson was out of town for a few weeks. Even though the groundskeeper tried to call the number that Nelson had given him several times, the supplier never came to fix the pump, since he didn't understand the groundskeeper, who didn't speak English very well, so he didn't know to make the necessary repairs.[32]

Yet when Nelson tried to explain what happened to the father of Genevieve and the Harrington brothers, they urged him to keep going despite the capsule heating up. So he decided to keep the capsule operating for as long as he could by continually refilling it with liquid nitrogen. He didn't even mention the one-week failure to other CSC members, even though he knew the one week of hot temperatures was fatal for the three patients in the capsule, because once they warmed up, decomposition would set in.[33]

Finally, by the fall of 1977, Nelson felt broken by everything that had gone wrong. He had gotten no money from the Harrington brothers or from Pauline Mandell for preserving her son Steven or husband, and he had received only sporadic payments of $50 or $100 from Guy de la Poterie for his daughter Genevieve. He had put almost half of his salary into the vaults. But for what? He felt the vault was no longer a joy or part of his great legacy, and he had run out of money, energy, and the expectation that the CSC could be saved. Moreover, he was broken and impoverished.

Yet, even after that, things got worse, as there were further repercussions from the failures in these early days of cryonics. For example, around March 1979, Nelson opened up the Mandell capsule, and as their parents requested, he removed both children and delivered them to Klockgether's mortuary for a service and convention burial. Since the children's parents no longer believed in the potential of cryonics, they felt it was time to acknowledge their children's death.[34]

At the same time he made other arrangements for three of the patients—Mildred Harrington, Steven Mandell, and Pedro Ladesma, who were transferred into conventional metal boxes, while he left four others, Louis Nisco, Marie Sweet, Russ Stanley, and Helen Kline, sealed in their capsule.[35] So presumably Nelson was ready to move on.

Cryonics in Court

But then came the lawsuits. The first one was from the Halpert family, who sued Joseph Klockgether and Nelson for not taking control of their mother's storage and maintenance at the CSC vault, even though they had not met any of the family and had nothing to do with freezing or storing their mother. They sought to portray Nelson as a despicable con man, using Klockgether as his cover, and argued that the two had swindled fortunes from their victims. Their lawyer, Michael Worthington, planned to call on the Harrington brothers as their star witnesses, and he sought $2.5 million for breach of contract and $10 million in punitive damages, even though Nelson obviously had no such funds. Making matters worse, even though Nelson had a signed Anatomical Gift Act documents showing a donation of the body of their mother Mildred, the Harrington brothers claimed they had a verbal contract that Nelson would provide a $20,000 capsule and would replace the liquid nitrogen forever. Nelson had to come up with another $12,000 for his legal defense, after giving an attorney referred by the Orange County Legal Aid Society a $3,000 deposit.[36] To pay the rest, he had to sell his car.

At the trial, Nelson's attorney argued that Nelson based his practices on the possibilities proposed by top scientists, including cryonics' founder, Professor Robert Ettinger, and he used procedures developed by well-respected medical doctors, even though he told prospective patients that cryonics was still a "long shot" with no guarantees, which is why donations were necessary to use these new procedures. His attorney also pointed out that the Harringtons never donated $100–$300 monthly to help with the liquid nitrogen charges. Nelson and the CSC kept their promise by freezing Mildred, shipping her to California, and maintaining her in dry ice and then in a capsule, which they maintained for over three years. Unfortunately the judge wouldn't allow Nelson's lawyer to introduce the Anatomical Gift Act into evidence, which undercut Nelson's whole defense. In fact, Terry Harrington testified that he had never heard about the medical donor requirement to become a cryonic suspension patient and that he had no idea what he signed because he was so confused and distraught, claiming that he would never donate his mother's body to anyone.[37]

In the end things did not go well for Nelson. The judge seemed especially skeptical about the payments for all the expenses of people frozen in different locations and the cost of pumps, liquid nitrogen, dry ice, and traveling each day to the vaults and the temporary storage facilities. He also had concerns about the many capsules breaking down and Nelson's flights around the country. The judge also questioned why Nelson would go through all the trouble and expense on behalf of the Harrington brothers, if they promised to make donations each month but never made one. So he was clearly skeptical about cryonics and the operations of the Cryonics Society of California.

After that, even though Nelson's attorney Winterbotham presented witnesses who spoke of Nelson's dedication and his use of his own money to pay for cryonics expenses, the attacks of the opposing attorney, Thomas Northern, the newly appointed plaintiff's cocounsel with a smooth, charming appeal, were effective. It didn't matter that the defense witnesses, such as Sandra Stanley, enthusiastically supported Nelson's commitment and dedication to the cryonics cause.[38]

In the end the decision came down to what happened after the pump stopped in the capsule: the people inside it were lost forever. Moreover, Nelson's final defense witness, Frank Farrell, an engineer, who helped him in replacing dry ice, pumping water out of the facility, and moving bodies, ultimately proved beneficial to the plaintiff's case against Nelson, when he described how the capsule became less and less efficient and finally failed after Nelson ran out of the money while continually trying to fill it.[39]

Unfortunately the trial was like the nail in the coffin for Nelson as well as a dire perspective on the hope for the success of cryonics as a science. Northern, speaking for the plaintiff, summarized his case by describing Nelson to the jurors as someone who took advantages of grief-stricken families with "promises to take away their pain and loss." Worse, he described Nelson as someone who was starting a new kind of religion, where he was the leader with the power to decide who should live or die, and people had to pay to live.[40]

It was certainly a damning indictment, based on the perspective that cryonics was a sham science that had no credibility with medical doctors. As the plaintiff's attorney argued, both Nelson and the so-called science itself were frauds, as evidenced by the failure of the capsule resulting in the disintegration and permanent death of Mildred Harrington and the other individuals in the capsule with her.

In response, Nelson could only point to a verbal agreement, documents signed by the brothers, the history of the CSC and CSNY, and CCS Michigan's claimed policy of never performing a cryonic suspension without the donation of a body after death. His lawyer argued that Nelson used his good faith efforts to keep the bodies in cryonics suspension, using all of the funds they contributed to pay for their suspension and dry ice replacement.[41]

But in the end these arguments weren't enough, and the jurors found for the plaintiff against both Joseph Klockgether and Nelson. They owed the Harrington brothers $400,000; after an appeal and some legal maneuvering, that amount was reduced to $65,000 in a settlement. The money ended up going to the attorneys while Nelson got nothing. After that, on signing the settlement papers in 1983, Nelson decided to walk away from cryonics forever; he felt that all of his love and passion was gone.[42]

Nelson left the history of cryonics for almost two decades and returned to his repair business. He gave a brief interview to the historian for the Alcor Foundation, one of the new cryonics organizations that carried the original vision to create a viable business. Then he decided to write his life story, which turned into a 2014 book that may be turned into a major motion picture. Meanwhile, the continuation of the cryonics story leading to what cryonics has become today was left to others who have developed new high-tech solutions to make the potential of freezing a realistic possibility for many.

I have presented the history of cryonics to illustrate the difficulties that any new science or high-tech venture has in getting off the ground. As Nelson's story so poignantly illustrates, in the early stages, a technology isn't perfected and there are many defeats and scoffers. Critics may look on these early failures as an indication that the underlying scientific theory is flawed, and they may charge that any scientific researchers or individuals attempting to create practical applications and commercialize this research are not only flawed hopeless dreamers but, worse, con artists attempting to play to the gullibility of people to get their money.

That's what happened with Nelson's trial. Although Nelson had devoted his life work to spreading the hope of cryonics around the United States and had gotten about a dozen patients to be frozen using his equipment and methods, in the end the failures of the technology spelled the doom of his dreams. Perhaps without the lawsuit and court trial, Nelson might have continued, struggling to raise money and attract more patients to be frozen. But the loss at the trial pointed out what is often a failing of the first developers. They may be inventors and dreamers but they lack the business sense to build on that dream to create the necessary technology and organizational skills to create a viable organization. Then it is left to the next generation of researchers and business people who see the potential of the early vision and build on this to create a successful business. In this case, that role was played by Alcor and other organizations that saw the potential of cryonics with the improvements in the technology for freezing patients, and they had the funding and organizational skills to make that happen. So the next chapter of this movement belongs to them, since cryonics is now considered one of the major routes to the potential of future immortality. In fact, it is one of the few approaches to create a community of believers in its potential path to living forever.

In Part II of the book, I will discuss these other approaches, focusing on the ones that have the most potential and support in the scientific community and among the growing numbers of individuals who are seeking immortality or at least a longer life through one or more of these paths today.

PART II

The Major Approaches Today

CHAPTER FOUR

The Successes and Struggles of Cryonics Today

As one of the pioneers in the search for immortality, cryonics has had the rockiest road of all of the scientific efforts to turn back the clock of death. After Nelson's struggles in trying to fulfill the early hopes, his venture ended with the meltdown of some bodies that were frozen in Chatsworth, California, and a ruinous lawsuit that he lost. The successor companies then had their own struggles. As a result, the cryonics approach has been hobbled by controversies, scandals, and skepticism from the scientific-medical community.

Yet a few companies survived the struggle—the Alcor Corporation, the biggest survivor, based in Scottsdale, Arizona; the Cryonics Institute (CI), started by Robert Ettinger and based in Michigan; and KrioRus, based in Russia. They have continued to gain adherents, averaging one to two cryonic suspensions a month. A fourth organization, the American Cryonics Society in California does take on patients but doesn't have its own storage facility and relies on the CI in Michigan to handle that.[1] The two biggest companies, Alcor and CI, are organized as nonprofits, and each has about 150 "patients" in suspension. Alcor has about 1,500 members who have signed up for being suspended when they die; CI has about half of that number; and KrioRus has only about 50 patients.[2]

Besides these members, a fervent community of believers has continued to support these programs, believing that they and others can later be revived once the technology for reanimation has developed to make that possible. They also believe that future generations will want to reanimate them. Many of these supporters might like to be members, but they can't afford the annual membership fee and the cost for storing their body or just their head, and it is hard to get insurance to cover these substantial costs. For Alcor members, the annual membership fee is about $700, and the cost of a full-body storage is $200,000. For those who opt to only have their head stored through "neuropreservation" (since the brain is the seat of the individual), the cost is only $80,000. These costs include a transport fee and a treatment, storage, and revival fee. About half of the fee funds a "Patient Care Trust," which is a patient protection fund, so that the patients won't be affected should the company suffer a financial crisis. By contrast, the costs for CI are considerably less. Their annual fee is only $120, and it costs $1,250 for a lifetime membership. Once a patient dies, the cost for treatment is $35,000 (only $28,000 for lifetime members) plus the cost of transporting a patient to the facility if he or she dies outside of the area ($95,000, or $88,000 for lifetime members). This transportation is arranged through one of CI's partners, such as a mortuary that handles the initial freezing so the body can remain cool while being taken to CI's Michigan facilities.

Following is an overview of the development of cryonics as it expanded after the Nelson debacle to become the movement it is today.

The Basic Promise and Procedures of Cryonics

As has been the case, from the founding of cryonics in the 1960s after the publication of Ettinger's book, cryonics is fueled by the hope that if a body is frozen in time between legal and total death, where the body deteriorates, it will be possible to reanimate the body and restore the person to health through future medical procedures, or at least the procedure can revive the brain and provide it with a new healthy body. To this end, the cryonics companies arrange for a team to quickly get to the hospital (or wherever the person has died) as soon as possible after he or she has been deemed legally or clinically dead. This rapid response is critical because right after that the bodily processes begin to deteriorate. Once death occurs, if the team gets the body soon enough so that reanimation is possible, the team takes the body and puts it in dry ice or in an icewater bath to start cooling the body. At the same time, a mechanical respirator is used to restart circulation in order to maintain normal bodily processes.

In many cases, rather than waiting for the call that a person has just died, the team may be assembled while the patient lies dying with only a few hours left to live. This way the team can be there at the instant of death, because each passing minute, hour, or day after death means that the preservation of the body will be increasingly reduced in quality.

Unfortunately, after a certain point of deterioration, any hope of reanimation becomes impossible. Moreover, the ability to use cryonics depends on a patient experiencing a "good" death, where the individual's head and body are still intact, so a quick response can be effective. By contrast, if a person has died due to some traumatic experience, such as a beating, severe head injury, or fatal car accident, the person's head or body may be too damaged for any hope of reanimation in the future.

In any event, assuming the person has died where cryopreservation is possible, the team has to make the necessary arrangements to obtain the body. Among other things, these arrangements require having the necessary paperwork, such as a death certificate and an agreement signed by the patient or a proxy before death assigning the patient's body to the cryonics organization. In some cases team members have to persuade the hospital staff that the team is properly authorized to take the body. Should such a hang-up occur, this time delay can jeopardize the quality of the preservation of the body, since it can deteriorate as discussions drag on.

Once the patient is placed in an ice bath and the mechanical respirator restarts circulation, unless deterioration has gone too far the next step is to administer all kinds of medications to reduce the metabolism and stave off any further deterioration that starts once the body stops functioning. The goal at this stage is not to freeze the body but to reduce the temperature to slightly above the freezing point of water so that the body can be kept in this state to prepare it for transportation to the cryonics facility for the suspension operation. In some cases the body may have to be transported from the hospital to a place where these procedures can be done, such as a mortuary, if the place where the person dies isn't suitable or agreeable to these procedures.

Regrettably this ideal scenario may not always work as planned, resulting in delays and damages to the body. For example, as pointed out in "Generation Cryo: Fighting Death in the Frozen Unknown," sometimes there can be "catastrophically long delays."[3] One reason for these delays is that disapproving family members may intentionally fail to alert Alcor or other cryonics facilities that the person has died, even though the recently deceased has opted to be placed in cryonics suspension. Another problem can be that a patient is subjected to an autopsy, which will seriously compromise a cryopreservation because of damage to the body when it is out in the open. Then, too, if the patient suffered from an aggressive brain tumor or neurogenerative disorder, this could erase memories or aspects of the personality, and these cannot be restored, even with the much improved medical technologies of the future.

Assuming the patient's condition is suitable for suspension, he or she is rushed to the treatment facility for a series of surgical procedures. Among other things, the surgeon drills small holes into the skull to make sure the blood is continuing to circulate before it is replaced by a high-tech gel. Then, if the surgeons are preserving the whole body, they connect all of the major blood vessels of the heart to a device that lowers the patient's body temperature to a few degrees above the freezing point of water. They also connect the patient to a perfusion machine, which infuses a number of chemicals into the body. As "Generation Cyro" explains: "The idea is to wash out the body's blood and other fluids as quickly as possible, and replace them with a cryoprotectant. This high-tech gel is gradually added to the body to prevent ice crystal formation . . . (which) are like tiny knives that stab away at cells, disfiguring their shape and disrupting the connections required for normal organ function."[4]

Once the body temperature is lowered gradually to a liquid nitrogen temperature of –196°C, this gel causes the patient to become "vitrified." That means the patient's blood has been "transformed into a glass-like state, and free of ice crystals."

If only the head is being frozen, it is surgically removed from the body, which in some cases is preserved separately or in other cases is discarded. Then the head is placed upside down in a holding chamber, where all fluids are removed and the cryopreservation solution is added.

After these procedures, the body or head is prepared to go into a large tank called a "dewar" for permanent storage. At Alcor, these large tanks are filled with up to four whole bodies and five heads, called "neuros." These whole bodies are lowered head down into the tank while the neuros are placed in a vertical row in the center of the cylinder. Next, liquid nitrogen is slowly added so that the body and neuros are gradually brought down to the desired low temperature. The reason for placing the bodies head down is that if there is any loss of liquid nitrogen, such as if there is a leak in the tank and the nitrogen heats up and evaporates, the head will remain cold for as long as possible.

Once these procedures are done and the head and bodies are in place, the facility has to maintain the bodies by adding liquid nitrogen regularly. Usually this task is done once a week, although it might be possible to maintain the bodies without any further replenishment of the fluid for much longer—even up to six to eight months.

Ideally that's how the process should go, but all kinds of things can potentially go wrong, such as when Alcor experienced a number of setbacks—and even an accusation of murder for speeding up the death of one patient, Dora Kent, by inducing her death chemically before she was legally dead.

Behind the Scenes at Alcor

Although Alcor seems to have overcome its early challenges and scandals to become the largest cryonics company today, its early struggles were much like the ones that Nelson faced in creating the first facility to freeze patients. Alcor's big test of both the validity of cryonics and the company came in 1988 with the case of Dora Kent, its ninth patient, when members of the Alcor staff were accused of murder for inducing her death before she died from a terminal illness in order to facilitate her cryonic suspension.

The case began in December 1987, when Dora, the 83-year-old mother of Saul Kent, an early advocate for cryonics and a board member of Alcor, was at a Riverside nursing home, where she was suffering from Alzheimer's and pneumonia. She had been ailing for many years when she came down with pneumonia and death seemed imminent. While it is common for a standby team to go to the location of a dying patient to perform the necessary cool-down operations and then transport the patient to Alcor for the perfusion and suspension process, in this case, the team decided to bring her to the facility before her death. Even though this was a medically sound decision, it became the basis of the criminal charges that threatened to destroy Alcor with potential repercussions for the whole cryonics community, as Michael Perry describes in his notes about the Dora Kent Crisis.[5]

According to Perry, Dora previously indicated that she wanted to be suspended, and the neurosuspension of her head was fairly routine. In fact, the procedure was facilitated because there was no waiting time. Dora was already at the Alcor facility at the moment of death. However, since a doctor hadn't been at the procedure and therefore couldn't record a death certificate indicating the moment of death, the coroner raised questions about Kent's death. Soon the story became a media sensation. The big question played up in the media was if Kent's death was really murder and could the coroner get Alcor to release the head so that an autopsy could be performed. Alcor naturally resisted this request, since it would undermine the process of suspending the head safely and securely so that it might later be reanimated.

Initially the operation proceeded as normal. Mrs. Kent was in the last stages of dying, after she had experienced mental deterioration for two or three years at the nursing home. Now she was on oxygen, was no longer fed through tubes, and was weak and feeble. Then, on December 10, after a suspension team took her off the oxygen bottle and shaved her head to prepare her for the operation, she stopped breathing after an effort to resuscitate her. After her heart rate stopped, the team began the protocol for freezing, which included cutting open her chest, massaging her heart, maintaining the oxygenation of the tissues, and adding glycerol to her circulatory system. Eventually some team members placed Mrs. Kent's head in a nitrogen-cooled Dewar flask.

So far so good. Unfortunately, on December 14, things began to unravel, because no physician was present when she died to sign the death certificate. As a result, Mrs. Kent's death certificate wasn't accepted by the Bureau of Vital Statistics, and her death had to be referred to the coroner.

The result was a series of visits and raids starting on December 15, while Alcor team members packed up and took away a load of the more critical files. But eventually, on the following day, the Riverside County coroner's office took Kent's headless body and performed an autopsy on it.

Initially the coroner determined that the cause of death was pneumonia. But later the autopsy revealed the presence of certain metabolites in the body, which suggested that she was still alive at the time of her suspension, although Perry and the team asserted that these drugs were given after her death. The coroner then sought the head for an autopsy, along with all of Alcor's patient records and bodies. But Alcor workers refused to produce the head or surrender other patients' bodies, leading to the arrest of several Alcor workers and volunteers though none were charged. A week later a SWAT team raided the Alcor facilities and seized most of Alcor's property, though by December 26 they returned most of it. Needless to say, the press had a field day with the case, generally presenting a "horrible" picture of Alcor's activities. One story even proclaimed that Alcor "starved Mrs. Kent 6 days before cutting off her head."[6]

Meanwhile, the fate of Alcor and cryonics hung in the balance. For example, on December 28, the deputy coroner assigned to the case the team members that it could still file criminal charges, based on the Alcor team members using cardiac arrest instead of brain death as the criterion for deeming a patient to have died.

In a scene that could be out of a movie, on December 31 several Alcor workers moved Dora Kent's head from the facility to keep the coroner from obtaining it and conducting an autopsy, because that would likely end any chance for a successful reanimation. Then, on January 7, when a team of deputy coroners and other officers barged into Alcor with a search warrant for Mrs. Kent's head, they weren't able to find it; they took other documents and several Alcor workers to the police station for questioning for several hours. Meanwhile, Alcor's lawyers did what they could to get any charges dropped.

Finally things began to turn in Alcor's favor. At a January 13 court hearing, the court issued a temporary restraining order against destroying or damaging frozen human remains at Alcor, and eventually on February 1, a preliminary injunction was granted that protected Mrs. Kent and the other Alcor patients from autopsy. Eventually the California Supreme Court vindicated the Alcor team, so the prosecutor's attempt to establish a homicide was abandoned.

Later, when the prosecution attempted to charge the Alcor staff with practicing medicine without a license, the staff was vindicated, since it is hard to claim that someone is practicing medicine on a person deemed to be legally dead. Finally the case was settled out of court with a $90,000 settlement to the Alcor workers, along with the return of their confiscated property.

Eventually these struggles in the late 1980s helped provide a legal standing for all of the cryonics services that competed during the next three decades and led to Alcor emerging as the largest and strongest of all of them.

The Battle over the Fate of Ted Williams

While Alcor continued to grow through the 1990s, it faced other challenges, as the cryonics process of freezing individuals to reanimate them in the future remained extremely controversial. Though Alcor had won a legal battle enabling them to continue to freeze patients once they were declared dead, most of the world remained skeptical, which sometimes led to dramatic legal struggles.

This occurred most notably in the Ted Williams dispute that began in 2002 over whether his head and body should have been cryopreserved in the first place. Then in 2007, Larry Johnson published the tell-all book *Frozen* about misdeeds by Alcor staffers based on his experiences and information gained by working at Alcor in 2002. Alcor sued Johnson, his coauthor, and the publisher, and eventually won, while Johnson went bankrupt.

The Ted Williams controversy began when he died at the age of 83 in July 2002. Also known as "the Kid," Williams was already in the public eye as a famous left fielder for the Boston Red Sox. He was honored as a Baseball Hall of Famer, and some considered him the greatest hitter in the history of baseball. Supposedly in 1996, Williams amended his will to state that "he wanted to have his body cremated and scattered over his Florida Keys fishing grounds along with the ashes of his Dalmatian named Slugger who died in 1999."[7]

But his son, John Henry, a believer in cryonics since 1997, wanted to have his father frozen. Committed to cryonics as a way to achieve life after death, John Henry claimed his father had changed his mind and "warmed to the idea of cryogenics later in life," although the nurses who cared for Ted Williams in his later years said he never wavered from wanting to be cremated. Nevertheless, John Henry was determined and claimed he had gained power of attorney and health proxy for his father, so he had the power to make all the decisions for Williams. As a result, John Henry ignored his father's will and made arrangements to have his body preserved. Even though Ted Williams had never applied to have this procedure at Alcor, six hours before he died, John Henry faxed a completed application or scrawled note with his father's forged signature. John Henry claimed it was written by his father from his hospital bed, stating that he agreed to cryonics preservation. John Henry also gained the support of his younger sister Claudia Williams, but his half sister Bobby-Jo Farrell disagreed with going against their father's wishes. The result was a big legal battle, with Farrell insisting that Ted Williams's final wishes were to be cremated and his ashes scattered over the Florida Keys, not to have his head and body preserved. Moreover, Bobby-Jo accused John Henry and Claudia of freezing their father so they could preserve his DNA and sell it in the future.[8]

Despite the public ridicule that accompanied their decision to spend this huge sum for freezing their father's body, John Henry arranged to have Williams's body packed in an icebox and shipped to Alcor's Life Extension Foundation. This foundation offered three basic options—a whole-body procedure for $120,000, a $50,000 neuroprocedure that just froze the head, and a third procedure of freezing the head and body separately for about $100,000, which is what they chose. According to Claudia Williams, who wrote the book: *Ted Williams, My Father,* she and John Henry made this choice to freeze Williams not for money but for love, since he and his sister couldn't bear the thought of losing him forever. John Henry even showed how much he believed in the power of cryonics by agreeing to have his own body preserved in the same way as his father, which occurred when he died two years later of leukemia at the age of 35.

But as much as Bobby-Jo fought to have her father cremated, she had to abandon the legal struggle when she ran out of the funds to continue.[9] Along with her husband Mark Farrell, she spent close to $100,000 battling Williams's estate and John Henry Williams, and the money finally ran out. Ironically, their suit ended three months after John Henry died of leukemia and was stored at Alcor's facility.

For a time the controversies Alcor experienced seemed to be at an end, and it had the legal right to continue the cryonics procedures, but in 2009 another huge flap arose when Larry Johnson's book *Frozen* was published by Vanguard Press. Johnson claimed all kinds of inappropriate goings-on at Alcor, including the mistreatment of Ted Williams's remains. The book proved to be a sensation but it also led to a massive lawsuit against Johnson, his coauthor Scott Baldyga, and Vanguard Press. Eventually, though, after a few years of court battles, Johnson went bankrupt and withdrew some of his more sensational accusations, and the case was settled in Alcor's favor. But once again the controversy helped fan the perception of cryonics as a dubious practice with a cultlike following, even though Alcor continued to thrive, along with a few other cryonics facilities that drew less attention, perhaps since they were smaller and less well known.

An Expose of Alcor and Another Legal Battle

The big battle over Larry Johnson's explosive book *Frozen* erupted because it contained scandalous revelations about what he learned during his short employment at Alcor for eight months in 2003. In the last few months of employment, he began acting like an undercover operative, collecting documents and photographs and wearing a wire every day to reveal what he claimed were gross misdeeds by the oddball operators of the facility.

As Johnson describes it, he wrote his book because he felt compelled to be a whistleblower after he observed some "really dark and possibly illegal things" that were being done at Alcor. He felt he owed it to everyone—including the patients, their families, the surrounding community, and even the true believers hoping for another future life—to document these abuses. He claimed these abuses included experiments on animals, safety violations, and rumors of suspicious, premature deaths of several patients.[10]

Before he saw these terrible goings-on when he got the job at Alcor, he had worked as a paramedic. But in December of 2002, after 25 years of dealing with medical crises on an ambulance, he felt burnt out and began looking for a new job. In doing his interviews, he toured the Scottsdale facilities, and when he checked out Alcor's website, he was impressed by the company's list of scientific advisors and conference speakers, which included some MIT and UCLA professors, U.C. Berkeley and Cambridge research scientists, and NASA's Jet Propulsion Laboratory scientists and advisors.[11]

So he took the job, even though he noticed problems in the organization on his tour of the facilities, such as the fact that all of the drugs were expired, which he thought was odd if Alcor considered these patients to be alive. He also wondered why Alcor stored paralytics, which stop a patient from breathing within seconds of being administered, if the patients were already dead.[12] And he thought some of the employees looked like skinny zombies or patients who had just been reanimated.

Yet despite his reservations, Johnson took the job and assumed the title of the "Director of Clinical Services." His job would be to train Alcor volunteers on emergency medical procedures and assist their emergency response teams in transporting patients to the Alcor facility from the location where they died or where they were initially prepped. Additionally he had to help prepare the bodies, administer drugs, and assist in the washout procedure in which the patient's blood was replaced by Alcor's cryopreservant chemicals.[13]

On January 29, 2003, Johnson's job at Alcor began. After he met the CEO Jerry Lemler, Lemler gave him his first assignment; writing up a Patient Assessment Report to assess when the current 620 members might die—or in Alcor terms, "finish their first life cycles." Though Johnson was skeptical that anyone could predict when each member might die, based on considering such factors as their age and health, he did so anyway, since that was his job.

In the next chapters of his book, Johnson continues to highlight the dysfunction he observed at Alcor, as if to support his decision for turning undercover spy and whistleblower. He remarks on the continual infighting in the company and the "endless power struggle" over who was really in charge. He notes numerous examples of wrongdoing by company officers, such as having expired anesthetic and paralytic drugs, since there would be no reason to sedate a corpse or immobilize dead people.[14]

Johnson also continues to characterize the Alcorian officers as spooky, malevolent individuals, like zombies in a horror movie, yet he continues to work there as if he is gradually working toward his spy and whistleblower role of seeking to expose the organization.

Johnson further describes the Alcor leadership as a group of moneygrubbers and visionaries seeking to spread cryonics across the planet. Johnson also uses the book as a platform to bring up unsavory incidents from the past of the major players in Alcor. For example, he describes how Jerry Lemler, chosen as president because he had a medical degree, was previously in charge of an insane asylum, an apt background for being Alcor's president and CEO. Moreover, Lemler seemed to treat Alcor, a nonprofit corporation, like a source of funds for extravagant expenses, such as $400 lunches for himself and his friends.[15]

But what especially disturbed Johnson was hearing about the controversy surrounding Dora Kent, the mother of Saul Kent. According to Pratt, the investigating Riverside County coroner alleged that Alcor officers and employees had killed Dora Kent, so they could begin her cryosuspension more quickly because she didn't die soon enough. Supposedly, one Alcorian, Mike Darwin, killed her with an injection. That's why when detectives came with a search warrant to collect her head, the Alcorians told them the head was gone and no one would tell them where the head was. Without the head, Riverside authorities were unable to bring formal charges against the Alcor leaders, even though an earlier autopsy of Dora's headless body showed lethal levels of barbiturates in her bone marrow. Johnson also points out how the Alcorians were members of other radical scientific organizations, such as the Extropians and Venturists.[16]

The popular image of the Alcorians was certainly aligned with this image of them being weird eccentrics involved in dubious activities. So while the Alcor officials might want famous people as members and seek out publicity to attract potential members, they didn't want any close scrutiny of what they were doing. Thus they were very cautious in doing media interviews, had a checklist of things to look out for, and kept their answers brief.

Meanwhile, Johnson continued to document their bad behavior, and he criticized the celebratory atmosphere that accompanied a suspension. For example, after Dr. Thomas Munson, a longtime Alcor member, died, the Alcor team prepared his body using the usual cooling and suspension process. But soon after his body arrived and was wheeled into the OR room at Alcor, Johnson expressed his shock at what he observed: about a half-dozen Alcorian spectators were allowed to watch the procedure, and they turned what should have been a somber after-death affair into as a kind of party. According to Johnson, who was shocked by their attitude, the spectators were "chatting, joking, and patting each other on the back . . . It was like a tailgate party before a big football game. Even more amazing to me, many of them had brought cameras and were snapping their own souvenir pictures of Dr. Munson's corpse, taking turns posing around the body."[17]

Yet, for the true cryonics believers, such a celebration of "death" was perfectly fitting, since they were there helping their fellow member on his way to "a new life, maybe to immortality." So the event was like a "happy bon voyage" party. Why not celebrate?

But after the operation ended with Dr. Munson's head being placed in an isolation box, Johnson was skeptical that this method of preservation would really work, because with the extreme temperature drop to –321°F, the brains would crack and split apart. He questioned whether the doctors of the future could use advanced technology to fix so much damage.[18]

Then Johnson reported still another botched operation when an 88-year-old man was discovered by a friend, after he had been lying on the floor of his home for two days. He was then taken to California Hospital Medical Center, where he died, and now the Alcor team needed to pick him up. But there were all kinds of delays in getting the necessary paperwork for the hospital to release the body to Alcor. Then, after the patient was infused with the cryopreservant solution at Klockgether's mortuary, there were more delays in finding a surgeon to perform the decapitation, so a local veterinarian did it. Then there were several more delays in getting a death certificate. Meanwhile, the man's body warmed in the morning sun, and it took another four hours before the team was ready to transport him to Alcor. When they arrived in the late afternoon, the patient had been in the back of the minivan for 10 and a half hours, and during this time the ice had begun to melt. So all fluid inside the patient's body had seeped out and the body bag leaked, soaking the mini-van's interior with these fluids.[19] Thus the patient's chances of being reanimated in the future were greatly impaired.

It was one more episode leading Johnson to stay at Alcor to seek the necessary evidence to expose Alcor. For a time, he took notes about his growing disenchantment with the group, and he even sought to become a more trusted employee by becoming a member who signed up to be cryo-preserved when the time came.[20]

Johnson also describes witnessing his third cryonic suspension, this time involving an 82-year-old woman, dubbed A-1234. She was found dead in a Hollywood-area nursing home, but this time because of the delay in finding the body, the team brought her directly to Scottsdale rather than going to Klockgether's mortuary for the initial washout. Again, Johnson found several things wrong with the procedure that disturbed him. Not only did the body smell terribly after traveling through the desert in an unventilated and uncooled moving truck, but Alcor's retired surgeon Jose used a hammer and chisel to decapitate her head by hacking through her neck bone and finally wrenching off her head. Then he drilled a hole into her neck bone to insert a handle that the Alcor employees used to carry the heads around upside down. After that, one of the lab assistants, James Skies, disposed of a bucket of waste and hazardous chemicals from the operation into the city storm drain behind the building instead of in a toilet or sink. Johnson was further dismayed when Alcor officials ignored his complaints about dumping toxic chemicals and disease-ridden blood into the storm drain. In his view, such activities as the toxic dumping and the brutal cryosuspensions showed that Alcor officers, board members, and other high-ranking members were engaged in criminal activities, despite the organization's stated mission of preserving bodies to be successfully reanimated in the future.[21]

Yet, as appalled as Johnson was by what he observed, he still sought to get closer by joining Alcor as a member in order to be welcomed into the inner circle, all the better to expose the organization's misdeeds. Though it would take three months for his membership to become active, he felt that his fellow employees would know that he had begun the process and that would be good enough for them to accept him as one of them.

After that Johnson began to engage in all kinds of undercover work, such as going into the offices of Alcor officers at night to look at the member files to learn more about the different types of people who joined Alcor. In the course of investigating, he found records of the many donation campaigns in which Alcor repeatedly sought money from members and urged them to contribute on the grounds they could write off any donations and deductions on their taxes since Alcor was a nonprofit organization.[22]

Johnson also found that most of Alcor's current membership was made up mainly of sick people, including AIDS patients, cancer victims, and people with brain tumors, who desperately looked to Alcor to help them gain a second life. He thought that the Alcor CEO was engaging in misrepresentation by misleading prospective members to believe that this new future was certain, when it was only a hope, if future medical technologies made this possible. In Johnson's view Alcor's hard sell sounded like hucksters pitching expensive snake oil to the terminally ill. Worse, Lemler falsely claimed that certain high-profile celebrities had become Alcor members when they had not, such as naming Larry Flynn, the well-known publisher of *Hustler* as an Alcor member, when he had only briefly considered the possibility.[23]

Though Johnson wasn't there when the controversy erupted about Ted Williams's head and body being preserved at Alcor, he spoke to Alcor employees and found records about it, so his account became one of the most sensationalized parts of his book by the media. As Johnson describes it, John Henry took control of his father after Ted's stroke in 1994. Later, in 1999, John Henry came up with the idea of freezing his father for profit since he also planned to preserve his father's DNA and sell it to the highest bidder or to people who wanted to clone Williams.[24]

Then, while Williams's body was on an air ambulance to Scottsdale, John Henry signed the authorization for his father to undergo cryonic suspension, claiming he had the power of attorney to do so. According to other accounts, he forged his father's signature on some documents. In any case, once Williams's body arrived, the operation to prepare his body and remove his head to be stored separately were like a circus, and his dead body was mutilated and mistreated. The nonessential bystanders stood around flashing pictures and chatting merrily. Because of Ted's celebrity, the Alcor volunteers were in a photographic frenzy. Many brought in family members to pose around his dead body.[25]

After that, the operation proceeded like a scene from a *Frankenstein* movie. The two men who performed the operation were dressed as surgeons but weren't real doctors. Hugh Hixon was a mechanical engineer, and Mike Darwin was a dialysis machine technician. Then, according to Johnson's description, citing the OR log and the Alcorians he spoke to, there were problems in getting the cryoprotectant into Williams's blood stream fast enough, and when the decapitation was going too slowly for Darwin, he grabbed a carpenter's hammer and chisel to saw through Williams's neck. Unfortunately, just as Darwin finished sawing, he learned that this was supposed to be a full-body suspension, not a neuro-suspension.

Lemler called John Henry and convinced him to have his father's head preserved as a neuro, and Alcor would still store his body as well. Even more mistakes occurred when the perfusion tubes were accidentally knocked out so that Williams's blood poured out onto the floor. Then, instead of placing Williams's head into a head-freezing device, the safest way to cool it down to –321 degrees, the Alcor team placed his head in a CryoStar experimental cooling machine until it could be permanently placed into a Dewar tank. As a result, his brain ended up cracking inside the faulty CryoStar, which registered on Alcor's Crackphone "like earthquakes on a seismograph.[26]

After this, Johnson's book reads like a mystery thriller, as he describes how he stayed late at the office to look through the files of different officials, feeling like an "amateur undercover agent gathering evidence."[27] Among other things, he discovered that Alcor didn't have Ted Williams's signature on any documents, supporting Billy-Jo's claim that John Henry committed fraud, and Alcor went along with it to get this celebrity patient.

Johnson also began secretly recording conversations to show how the Alcorians used various threatening and underhanded tactics to get money from clients. He also taped a microphone to his chest and ran the cable under his shirt to his front pants pocket so that he could make recordings with Alcor's key officers. Later he called Bobby-Jo's lawyer, John Heer, to report some of the evidence he uncovered to support his client's claim against Alcor, such as not seeing Ted's signature on the suspension documents and membership agreement.[28]

So now, as Johnson describes it, he was living two lives to expose Alcor—by being an "Alcor employee by day" and "a whistle-blower at night"; and since this was scary, dangerous stuff, he began keeping a 9 mm Beretta in his truck, in case the Alcorians found out what he was doing and sought revenge.[29]

Despite the risks, Johnson kept digging, and he was especially interested in Alcor's misdeeds in the Ted Williams case, such as getting ready to quickly relocate Williams's head if they received a court order to release it, so they kept the malfunctioning Cryostar for almost a year. Otherwise, if the head was stored in a Dewar, it would be hard to get the head out quickly, but if it was in the Cryostar, they could take it away before any investigators or law enforcement officials could get it.[30]

Johnson also describes how he sought to cover up making copies of incriminatory documents to serve as evidence that would be admissible in the courtroom. For example, he persuaded Lemler to sign a letter giving him the permission to use various photographs or documents for publication by claiming he wanted to show this letter to other Alcorians so that they wouldn't be suspicious when he wanted to copy files. He even mapped out the locations of different cameras so that he could sneak into the building at night without being caught on camera.

This undercover operation went on for about two months as he searched for two main types of evidence—documentation of the environmental infractions he believed Alcor employees had committed, such as by dumping the remains after preparing a patient for a suspension, and evidences of Alcor's unlawful possession and mutilation of Ted Williams's body. Johnson began carrying his gun on him at work as well and gained even more inside information on how Darwin gave Dentinger an injection with a paralytic to speed up his death.[31]

Then, shortly before he left Alcor, in an account later featured in the major media, Johnson watched some Alcorians transfer Williams's head out of the CryoStar, now that the company officers felt secure no one was going to attempt to take the head away from them, even though the head suffered 16 cracks while it cooled down. Then the Alcorians transferred it into the Neuro Vault, even though one Alcorian, Hugh Hixon, had trouble dislodging the tuna can that propped up Williams's head. So after a few kicks that didn't work, he grabbed a monkey wrench and swung it. But instead of striking the tuna can, he hit Williams's head such that "tiny pieces of frozen head sprayed around the room."[32]

Shortly after that observation, Johnson's undercover efforts came to a screeching halt when the Alcorians discovered what he was doing. His unveiling began in July 2003, about eight months after he first began working at Alcor. At the end of July he contacted a *Sports Illustrated* writer, Tom Verducci, telling him that he hoped to expose Alcor quickly through the national media so that their misdeeds would be fully investigated. But in early August, a few days after he drove to Los Angeles to meet with an police detective to share his recordings about the suspicious deaths of Dora Kent and John Dentinger, the *Sports Illustrated* editorial and legal department contacted Alcor to do some fact checking. Verducci called shortly before the magazine planned to call Alcor to forewarn him of their plans.

Now Johnson knew it was time to leave Scottsdale, and the rest of the book describes how he and his wife were on the run, in fear of their lives. Their fear was very real, since as they were packing to leave, there was a loud banging on their door, accompanied by shouts, calling Johnson a traitor. "We'll get you . . . We'll kill you!" someone called out, before the shouts of neighbors scared away the man who was banging on their door.[33]

Meanwhile, in mid-August 2003, Johnson's publisher, Vanguard Press, pushed up the planned publication date of his book *Frozen*, so now the networks, radio stations, and newspaper reporters came calling, even from the international press in Russia, Japan, and other countries. However, Johnson decided to forgo the TV appearances, feeling that it was too dangerous, and he went into hiding; but before he fled, he did one interview with CNN and another on *Good Morning America*.[34]

Soon after that, Johnson and his wife fled Scottsdale. For a time they hid out in a guest house of some old friends from Glendale; in September, once news broke about how homicide detectives in Los Angeles were investigating Alcor's involvement in John Dentinger's death, Johnson did some more radio and news interviews.

Meanwhile, Alcor sued him, and he sued them back. As Johnson describes it, Alcor accused him of breaking corporate confidence, even though he had never signed a confidentiality or nondisclosure agreement. Alcor also accused him of disclosing company secrets, such as their "secret chemical cocktails," although whatever he had revealed was already posted on their website or featured in their widely distributed newsletters.[35]

In any event, after three months of lying low in Glendale, Johnson decided it was time to move on and find a job. But, according to Johnson, the Alcorians were still threatening him, while the lawsuits were proceeding very slowly. For instance, he received a typewritten note in the mail with no return address, warning him: "We know where you are . . . we will get you you will pay drop your campaign against cryonics!!!" Then, a few days letter he received another even more threatening letter that told him: "Consider yourself dead . . . You are a murderer and a trader[sic] your death will be slow."[36]

But nothing ever came of the threats, and Alcor's case against Johnson slogged through the courts.

Then, in an epilogue six years after he fled Alcor, Johnson provided a brief update on the major players discussed in the book. Most significantly, Bobby-Jo Ferrell reached a settlement with her brother John Henry so that she could no longer seek to release Williams's body from Alcor. For a time, her former lawyer John Heer continued the fight but ultimately gave up, so Williams's head still resides at Alcor. John Henry's head and body went there too, after he died of leukemia in March 2004. After Lemler resigned as president and CEO two weeks after the *Sports Illustrated* story, Saul Kent took over the organization. Johnson concludes the book by stating that at least 888 active Alcorians, some worth hundreds of millions and even billions of dollars, considered him a "mortal enemy," threatening their everlasting life.[37] Meanwhile, the organization continued to survive without any regulation. Even as he wrote the book in 2009, Johnson indicated that he continued to fear for his life, even though he felt he had done the right thing in exposing the organization. But it came at a price. As he concludes his book:

I don't think things will ever return to normal for my wife and me . . .
Doing the right thing is not always easy.
Beverly carries mace everywhere she goes.
I still carry my gun.[38]

The End of the Alcor Lawsuit

Whether or not Johnson was actually in physical danger from Alcor, as he claims in his book, Alcor quickly struck back in the courts, where it ultimately prevailed.

On its website, Alcor provides a blow-by-blow description of the legal battle from October 2, 2009, to May 19, 2014, in which it accused Johnson of defamation and other charges. Alcor especially denied the media accounts that it mistreated the remains of Williams, and it pointed out that Johnson was not employed at Alcor when Williams was cryopreserved. Moreover, on July 7, 2009, Alcor obtained a judgment in Maricopa County, Arizona, against Johnson, forbidding him from disclosing "any confidential or

illegally obtained information about Alcor, its members, or its operations."
Further, Alcor claimed in its lawsuit that the book was based upon "illegally obtained and misleading information that also violates the privacy rights of our members and employees." Even by Johnson's own admissions, during his employment for approximately seven months during 2003, he "stole confidential documents and medical photographs from Alcor and secretly recorded private conversations with Alcor staff members."[39]

Johnson took steps to evade the court process for a time, since he failed to appear in court in both New York and Arizona, where Alcor sued both him and Vanguard Press, and he repeatedly sought to avoid the service of process, even though he readily appeared on national TV to repeat his claims against Alcor. Meanwhile, in both court and in the media, Alcor sought to dispute Johnson's claims, such as his claim that he saw Alcor staff striking Williams's head with a wrench; multiple individuals who witnessed the patient transfer procedures stated that his claims are a complete fabrication.[40] Likewise they disputed his claim that Williams's head was stored in an unsafe, malfunctioning CryoStar freezer, and they disputed his "sensationalized" reference to an Alcorian using a hammer and chisel in cryopreservation.

At the same time, Johnson's ability to damage the reputation of the science of cryonics gained credibility because cryonics already had a suspect reputation for much of the public and the medical profession. So it was easy for Johnson's claims, whether valid or not, to contribute to a widespread view that cryonics didn't work and was the province of mostly wealthy dreamers who were willing to invest a portion of their substantial fortunes for the dubious promise that they could eventually return to life and achieve immortality.

Ultimately though, things did not go well for Johnson in the courts. On March 17, 2010, the New York Supreme Court granted Alcor's motion to amend its lawsuit against Johnson, his coauthor Scott Baldyga, and publisher Vanguard Press to include dozens of defamation claims, including Johnson's false allegation that Alcor mistreated Ted Williams. The Court also approved a motion based on the Arizona judgment against Johnson in place since July 7, 2009, that he should return all materials he took from Alcor, and prohibited him from making any statements against Alcor.

Then, on May 28, 2010, the Superior Court of the State of Arizona found Johnson in contempt of court and issued a warrant for his arrest. An associated judgment also ordered him to pay Alcor $34,100, in addition to what he owed from a previous July 2009 judgment since he violated a 2004 settlement agreement in which he agreed to make no further public statements about Alcor, after he was sued for distributing and misrepresenting confidential materials taken from Alcor in 2003. The court considered his "defamatory and fictionalized book *Frozen*" to be a violation of this prior settlement agreement and the July 2009 judgment. When Johnson failed to appear to respond to his transgressions, the court issued a judgment of contempt against him and a bench warrant for his arrest in any state where he might be found.

Soon after that, another arrest warrant was issued in his new home state of Nevada. After he posted bond to avoid jail in Nevada, he invoked the Fifth Amendment more than 300 times to avoid incriminating himself in a deposition on October 20, 2010. As a result, due to his failure to meaningfully cooperate with the orders of the court or answer questions, he was required to appear at another hearing on November 9, 2010, and pay Alcor for additional sanctions of over $40,000, plus $100 for each day he continued his conduct in contempt of court.

Finally, in February 2012, the litigation came to an end after Johnson filed for bankruptcy and made certain concessions in a public statement, acknowledging that some of his recollections and conclusions were incorrect and apologizing for these errors. In the end, Alcor felt victorious and vindicated. As Johnson stated in his public apology:

> When the book *Frozen* was written, I believed my conclusions to be correct. However . . . in light of this new information from Alcor, some parts of the book are questioned as to veracity.
>
> For example . . . I am not now certain that Ted Williams'[s] body was treated disrespectfully, or that any procedures were performed without authorization or conducted poorly.
>
> To the extent my recollections and conclusions were erroneous, and those recollections caused harm I apologize.[41]

Thus, for all intents and purposes, the litigation over Johnson's actions and the book were over. Aside from some final legal maneuverings, Alcor clearly won the case, thereby clearing its name from Johnson's derogatory claims.

Since then, Alcor has continued to grow, becoming the largest of the four major cryonics organizations in the world. There is still much skepticism from the medical community and from the general public, but believers continue to believe in the hope of cryonics to give them a renewed life in the future.

Alcor's Continued Success and Cyronics Today

Since the lawsuit against Johnson effectively ended, Alcor's statement on its website has highlighted the state of cryonics today. Despite its many difficulties, cryonics still holds out the dream of reanimation for everyone who has been or will be preserved.

Despite cryonics' early turmoil, Alcor has continued to grow and prosper. As of 2016, Alcor was over 40 years old, has about 1,000 members, and its Patient Care Trust Fund is endowed with over $9 million to provide for the long-term care of several hundred cryopatients. Alcor claims to have had positive membership growth almost every year since its inception, and it is now the largest cryonics organization in the world. It has about a dozen employees in Scottsdale and works with about 20 part-time independent contractors around the United States. Mostly they handle emergency standby and rescue efforts when a patient is about to die or has just died in order to provide care and transport to Scottsdale.

The company acknowledges that its major barrier to acceptance by mainstream medicine is that no patients have yet to be reanimated, so there is no proof yet that the concept works. So far it just offers a belief in the possibility of a new future, based on advances in medical technology, which the other paths to living longer and possibly forever are pursuing in different ways.[42]

At least Alcor freely acknowledges its major challenges. One is a lack of clear feedback on the results of its treatment, since the outcome of their procedures to preserve and return the dead to life won't be known for decades or even centuries. While cryonocists believe that future technologies will be capable of repairing cellular damage in suspended patients, they also recognize that if a patient experienced severe brain damage before they were cryopreserved, "repairs may be delayed, may be incomplete, or may be impossible." But cryonocists don't know whether the damage will be irreversible if a patient experiences a short period of warming after legal death.

Then, too, cryonocists are not able to develop objective criteria to compare one case with another because so many uncontrolled variables go into the preservation and storage process. For example, an airline delay or missed connection could result in a delay in getting a patient to Alcor, or a patient might experience several hours after death before being discovered, even though the transportation to Alcor goes smoothly. But perhaps the speed of preserving the patient once at Alcor might outweigh the damage due to the patient warming up before the Alcor staff can start the perfusion process.

Another problem cryonics has faced is the rift between cryobiology and cryonics, as the Society of Cryobiology members have been skeptical of the cryonics vision. As a result the society has actively discouraged scientists from doing any work that might advance cryonics and has even adopted a bylaw whereby they might expel any member who practices or promotes cryonics.[43]

Thus, even though the procedures practiced by Alcor have advanced far beyond the crude freezing methods imagined by most cryobiologists, and molecular nanotechnology experts have expressed strong support for the cryonicists' work, cryobiologists and other scientists have viewed cryonics negatively. Still, Alcor is hopeful that as more papers are published that describe technical advances, the cryobiologists and other scientists will come to recognize cryonics research as a legitimate specialty.

In turn, this medical and scientific hurdle is reflected in legal barriers. For instance, cryonics is not formally recognized in the laws of any state in the United States, although it is still legal and regulated by certain laws. For instance, cryonics facilities have to comply with state laws affecting the transport and the disposition of human remains, so Alcor makes arrangements with licensed morticians to meet these requirements. It also complies with the regulations of OSHA, the EPA, and other agencies.

Still, as Alcor notes, terminal patients have long had the freedom to choose, since the United States has a well-established cultural tradition of honoring the desires of terminal patients. Even so, Alcor and other cryonics organizations have faced numerous hurdles, as cryonics is not considered an accepted or recognized therapy in the general medical community. Medical professionals commonly view cryonics as an unusual anatomical donation at best, and some see it as fraud. Given these negative attitudes, some hospitals have delayed releasing patients to Alcor or they have forbidden Alcor staffers to use cryonics life support equipment or medications in their facilities. Commonly, though, when nurses and physicians learn that cryonics is a "sincere practice," they are willing to do what a patient wants or at least to not interfere with the patient's wishes. And, at times, medical staff members even help with some cryonics procedures, such as administering medications or performing chest compressions when legal death occurs and Alcor staff members aren't present at that time.

This lack of medical acceptance also means that patients who are remote from Alcor have to be moved to a mortuary for the blood-replacement procedure before they are transported to Alcor, even though performing these initial preparations in a hospital would be preferable. Still, Alcor and other cyronicists hope that in time they will have the same privileges as organizations that obtain organ donations from deceased patients.

Finally, another problem facing cryonicists is that in over half of the cases where patients want a cryonics suspension, legal death occurs before Alcor standby personnel can arrive. As a result, during this time these patients experience a warming that can cause severe cellular damage. An even greater danger is the threat of an autopsy in which the brain is dissected; anyone who is deemed legally dead after unexpected circumstances, such as an accident or homicide, is liable to be autopsied. Unfortunately this procedure largely destroys the patient's chance for a brain recovery later. That's why Alcor encourages patients to sign Religious Objection to Autopsy forms in the states where they can do so—California, Maryland, New Jersey, New York, and Ohio.

Yet in spite of all these hurdles, Alcor and the other major cryonics providers remain hopeful that one day they will be able to show that their procedures work so that they can gain medical acceptance. Though they acknowledge that the idea of immediately providing cardiopulmonary support after a cardiac arrest and cooling a body cannot be achieved in most cases, they continue to believe. Moreover, they claim that molecular records of memory persist in the brain, even hours after clinical death, so if the brain is the seat of the personality, the existence of this molecular memory provides still more hope for future reanimations. But ultimately only physicians of the future, wielding new medical technologies not yet known, can determine if the person who was frozen continues to exist.

So on that hope the future of cryonics patients rests, assuming they can cover the cost of their suspension—currently about $200,000 for a whole-body cryopreservation and $80,000 for a neurocryopreservation, plus additional nonmembership and maintenance fees.

And so, with some money and a lot of hope, the cryonicists look to the future that might be possible with the help of Alcor and the other cryonics facilities. Will that hope materialize? Possibly, if other high-tech, biotech, and medical research into prolonging life bear fruit and if one can live long enough to enjoy them or be preserved successfully after death. Then perhaps one might live on or live again. The next chapters deal with these different paths for achieving longevity or even defeating death.

CHAPTER FIVE

Living a Longer, Healthier Life

While cryonics may be a possible solution for those at the very end of their life, another approach to living forever is living longer until the new technologies make immortality possible. At least if one doesn't live long enough for that to happen, one will have the major benefit of a longer, healthier life. And now there is a science of living longer, which is commonly referred to as the life extension, longevity, or antiaging approach.

This approach to living longer goes beyond the health, wellness, and fitness approach of just eating better, getting regular medical checkups, getting regular exercise, taking vitamin pills and nutritional supplements, and the like. It also involves a comprehensive program based on scientific research of what one can do to live a longer, more vital active life.

There are even life extension doctors and an American Academy of Anti-Aging Medicine (A4M), founded in 1993, which is a nonprofit organization that promotes this field and trains and certifies physicians in this specialty. It actively engages in lobbying, public education, and public relations. As of 2011, about 26,000 practitioners in 110 countries obtained certificates, although many established medical organizations, such as the American Medical Association (AMA) and the American Board of Medical Specialists (ABMS), don't recognize this field. Even so, the organization continues to flourish and sponsors several conferences, including an annual World Congress on Anti-Aging Medicine, the latest held in April 2017 in Hollywood, Florida.

Another very active group is the Life Extension Foundation, founded in 1980, which has been dedicated to finding new scientific methods to eradicate old age, disease, and death. It has taken the lead in recommending many drugs and nutritional supplements, including high doses of antioxidant vitamins; remedies for arthritis, atherosclerosis, and Alzheimer's disease; and research on stem cells, fish oil supplements, hormones, and enzymes. It also has a directory of antiaging doctors and health practitioners around the world who are open to alternatives to allopathic medicine, including prevention, nutrition, and longevity practices.

Some antiaging methods recommended by the A4M and other antiaging practitioners, such as exercise and a healthy diet, are widely supported by established medical doctors and health practitioners; other methods, such as hormone treatments, are not supported by many in the medical community or by many scientists studying aging. Moreover, critics feel the A4M and many antiaging doctors and companies have engaged in misleading marketing practices to sell customers expensive products that are ineffective, resulting in some legal and professional disputes. Then, too, in the name of antiaging, some marketers have used deceptive practices to appeal to customers who want to live longer, healthier lives with assorted antiaging supplements, creams, and lotions. I even wrote about several companies that marketed Beau Derma skin cream and Revita Eye to combat aging by suggesting that they had endorsements from celebrities and a TV medical doctor to promote their products. But this marketing was designed largely as a scam to get credit card information and charge several hundred dollars for each victim, as I described in my book *Scammed*.

So what are the recommendations for living longer, based on a number of scientific studies that the life extension doctors use to support their methods? In this chapter I'll describe their findings. Many commonly accepted practices to live longer involve adopting a healthier lifestyle based on better nutrition, exercise, and avoiding life shortening practices such as smoking, abusing alcohol and drugs, and engaging in high-risk sports. However, some recommendations are still subject to dispute (such as taking certain nutritional supplements, vitamins, and hormones) like many new technologies in the antiaging field that have encountered skepticism since these practices are not yet accepted by the mainstream.

The Major Strategies for Living Longer

Some of these strategies for living longer have long been promulgated as a way to live a healthier life, though they haven't necessarily been linked to longevity and antiaging. But whether linked or not, virtually anything that promotes health, wellness, and fitness will contribute to a longer life, because one will be stronger, fitter, and better able to stave off disease or heal from any injury.

The main strategies include eating a better diet, exercising regularly, living a healthier lifestyle, managing stress, and having a positive attitude. The two strategies that have been most controversial are the recommendations to take certain supplements and some of the practices of antiaging medicine. In *Life Extension Express*,[1] David A. Kekich, the founder of the Maximum Life Foundation, has written a book widely praised by those in the antiaging community.

As Kekich writes in the introduction to his book proposing seven steps to living longer, the antiaging movement has gained its impetus from the advances in biotechnology. Before we were living in a time when there was no way to defeat aging. But now all that is changed, because we are living in an era "where advancing biotechnology will give us the tools to overcome aging. All the old attitudes are no longer relevant."[2]

To this end Kekich has written his book summarizing today's knowledge, technology, products, and services to help readers be alive and healthy when the age-reversal technologies of the future are available. Some of these products and services have come under scrutiny and criticism by the critics who question some of the practices of the antiaging movement. But rather than take sides where there is controversy, I'll report on their claims; then you can decide for yourself on the validity of the approach. Or maybe continuing research in this field will show where the antiaging recommendations are right or wrong.

The New Sciences Fueling the Antiaging/Longevity Movement

I'll describe these sciences in more detail in the chapters devoted to the major advances in biology, biotechnology, and technology. Here I want to point out how these life extension strategies have been based on this scientific research.

One key principle of the antiaging/longevity movement is that most of the killer diseases like cancer, heart disease, and stroke are age related. It makes sense, then, to go after the root cause of these diseases— aging— to avoid or cure these diseases. Therefore, instead of spending hundreds of billions of dollars on fighting a particular disease, it makes sense for researchers to find a way to overcome the ravages of aging itself.

These battles to defeat aging or at least slow it down are the result of the advances in biotechnology and information technology. Among other things, as Kekich points out, scientists can now turn destructive genes off; they can add new genes to patients using gene therapy; and they now know how to turn proteins and enzymes on and off to slow down or overcome a disease. As a result, disease and aging processes can literally be turned into programs that can model, simulate, and change the course of diseases and the aging process. And if we can do this now, just think how much more we can do with even more powerful technologies in the near future. A key reason is that some scientists now compare the biological processes to information processing. They think of the approximately 25,000 genes in each cell as akin to a software program. Kekich suggests that we are now making exponential gains in modeling and simulating the information processes that cracking the genome code has unlocked. Moreover, Kekich and others in the antiaging community believe that advances in biology will happen even more quickly in biology due to the "Law of Accelerating Returns," whereby each scientific development opens up still more opportunities.[3]

Currently the four top causes of death linked to aging that the antiaging supporters are seeking to overcome are heart disease, cancer, stroke, and arterial blockages. Other causes of death where aging can contribute to an inability to recover include diabetes, influenza and pneumonia, nephritis and nephrosis, septicemia, Alzheimer's disease, and unintentional injuries. Out of the top 15 causes of death, the only ones not related to aging are accidents, suicide, murder, and HIV. So overall, as Kekich notes, 85 percent of all deaths are due to aging, which leads to the degeneration of different organ systems in the body.[4]

A good way to think of the aging process is to imagine the body as a machine whose parts wear out at different times. The result is the machine may slow down in its operations or be unable to perform some operations but it still keeps going. But then it suffers a serious challenge or impediment and has to stop unless it can be fixed quickly; if not, it has come to the end of its life.

A number of these sources of breakdowns include:

- Mutations to the DNA occur when the cells divide, and over time these mutations or mistakes add up.
- Certain damaged proteins and other waste products can accumulate in one's cells and they are not efficiently excreted from the cells or one's body.
- Oxidation causes damage to one's cell membranes, DNA, enzymes, and mitochondria, which are sources of power in each cell.

Such damage to the cells can occur due to the interaction of various molecules and chemical reactions on the cellular level. A popular explanation is that this damage commonly occurs because free radicals cause oxidative damage to the antioxidant systems that are designed to clean up or reduce the reactivity of the highly reactive molecules. These molecules are normally contained in the mitochondria that help convert the food one eats into energy that the cells can use. Unfortunately, as aging occurs or one experiences a high level of stress or exercises too much, the systems in the cell can't keep up with the reactive oxygen that is produced. As a result, these molecules escape from the mitochondria and can damage nearby tissues. Damage can also occur when the immune system cells release reactive oxygen to destroy any invaders, such as viruses and other infectious toxins that cause disease. If too much oxygen is released, the inflammation that occurs destroys one's own tissues, not just the invaders.

Some of these details about how different genes, chemicals, and cells interact within the body can become very complicated and hard to understand, especially for those who don't have a grounding in these subjects. But the basic idea is that certain biological processes slow down, stop, or speed up and variously cause damage to one or more organs in the body. Further cellular damage can then occur because these processes can be negatively affected by external sources of damage such as smoking, alcohol, air and water pollution, and exposure to other toxins. It doesn't matter whether we ingest these sources of damage (such as eating foods with sugar) or we are exposed to a toxic environment. Over time, these toxins build up and eventually the body's ability to fight back against them is diminished until it gives up.[5]

Thus the strategies for living longer include ways to beat back these problems. You can also take other steps to prolong your life by avoiding some of the risky behaviors or places that might lead to an early demise. For example, driving too fast could lead to an accident; getting into a fight with a rage-fueled motorist could end in a shooting; playing unsafely with fireworks could lead to an explosion; or going into an unsafe place late at night could lead to a mugging. In other words, antiaging efforts can't help if one engages in certain dangerous activities, just as cryonics might not be able to offer hope for a later reanimation if someone's head or body has been severely damaged. But otherwise, for most people, these antiaging strategies can contribute to a longer life and maybe even to living indefinitely because of the potential for biotechnology to reverse, even permanently, some of the damages due to aging. Among those techniques that I'll discuss further in a subsequent chapter are:

- *Repairing the major types of cell damage that occurs in aging,* which can lead to heart disease, arthritis, diabetes, cancer, and other diseases. This is the approach of Dr. Aubrey de Gray, who has created the SENS strategy (Strategies for Engineered Negligible Senescence). This method is designed to repair or prevent the accumulating damage in cells and has been working to show the effectiveness of this technique by extending the life of elderly mice.[6]

- *Replacing your least viable stem cells with the best stem cells,* since all the cells of your body, including your stem cells, experience damage to their DNA. Usually this occurs due to the attack of free radicals created by the mitochondria in your cells and exposure to toxins in the environment. Unfortunately, using your own stem cells to replace damaged or aged tissue doesn't work very well, as you just get more of your own cells. But now some researchers have been able to turn ordinary adult skin cells into embryonic stem cells in order to repair any injured, diseased, or aged tissue.

- *Finding drugs with antiaging properties,* as these can act directly on those genes that contribute to living longer. To this end, researchers have been looking at long-lived and normal-living fruit flies, which have about 75 percent of the same genes as humans, to see which genes are different so that they can identify the genes related to living longer. So far they have found 400 such genes, and they have begun creating diagnostic tests and drugs to apply these findings to humans.

- *Using macromolecules in DNA, RNA, and proteins to treat and prevent disease by making changes at the cellular level.* These approaches include developing therapies to treat cancer and infectious diseases and create vaccines; developing nanotechnologies that work at the molecular level or even at the atomic level to regenerate human tissues; and preventing the telomeres at the ends of the chromosomes in cells for shorten or even lengthening them.[7]

- *Engaging in genome reengineering by using RNA interference* to put a bits of RNA inside a cell to turn off selected genes associated with aging. A goal of this approach is to rewrite the coding in the human genome to make one immune to aging damage, though this is a time-consuming process that can take many thousands of hours of detailed coding work.

- *Using nanomedicine to repair almost every cell in your body, atom by atom* to cure any damage due to aging, injuries, or disease.

- *Improving brain functioning by creating a direct link between the brain and computers* can help increase memory, learning, and cognition.[8]

Given all of these biological and technological developments, there is increasing hope that individuals living today will have much longer lives. The human life span has already been increased by about two months a year for the last five years, and within the next 15 years medical technology might be expected to add over one year to your life span every year. This development is what antiaging experts call each person's Longevity Escape Velocity, and presumably, as more and more medical developments are added to the mix, this velocity will increase.[9] While this is just a future projection, given the many streams of progress on different medical and biotechnology fronts, it certainly seems possible.

If you can keep living longer using various strategies to do so, you might not only live a longer, healthy life but do so forever. Then, of course, there is always the hope that cryonics could bring you back in this brave new future of immortals.

What are these strategies for living longer? The antiaging advocates propose a number of different steps you can take right now to extend your life.

One group of strategies involves eating right, which includes eating less; having a good diet; taking nutritional supplements to get the necessary vitamins, minerals, and other chemicals not in your food; and keeping your weight down. The other group of strategies involves living right, which includes getting plenty of exercise, having a healthy lifestyle, managing stress, learning to relax, having a positive attitude, and reducing your exposure to the toxins in the environment. Then, too, some antiaging medicines can help you now.

CHAPTER SIX

Eating Right to Live Longer

One of the major strategies for living longer is simply eating right, which includes having a nutritious diet and keeping your weight down. This may sound like what any health practitioner might advise. Numerous individuals and companies are selling healthy-living food products for people seeking to lose weight, resulting in all kinds of low-calorie and diet products sold in supermarkets. A popular TV show, *The Biggest Loser*, was devoted to weight loss, and TV doctors, like Dr. Oz, promote nutritional supplements for better health.

However, the approach of the antiaging and longevity advocates goes beyond promoting the goal of eating well and keeping weight low for improved health. Instead they draw on research that suggests that certain types of foods and nutritional supplements contribute to living longer and perhaps might enable some to live forever. The advocates also seem to be in general agreement about the types of foods and diet that are good for you, so I'll cite the most commonly mentioned ones here.

The Calorie-Restricted Diet

One example of this approach is the calorie-restricted (or CR) diet, as described by Joel D. Wallach and Ma Lan in *The Agebeaters and Their Universal Currency for Immortality* (or *Immortality* for short). In developing this diet, they combine their research representing 94 years of study from separate worlds—the United States and China—whereby they studied centenarians throughout the world to discover what was common to all of them. In the course of this research, they studied the cultures, diets, health practices, diseases, and daily lives of individuals from China, the United States, and 20 other countries They also conducted 20,000 autopsies—3,000 on humans and 17,500 on more than 400 species of captive wild zoo animals. They concluded that there are four main daily habits and activities that contribute to a long life, based on eating the right food, getting the right nutrients, and restricting one's diet, without suffering from malnutrition. They outline the specifics of this diet in nearly 500 pages. In their view, the medical approach to health and disease has cost Americans extensive misery, early death, and great deal of money, while the American way of life has left people suffering with cardiovascular disease, obesity, diabetes, arthritis, and other ills.[1]

One big problem is that when society shifted from using wood as the universal fuel to electricity, humans no longer had access to the traditional sources of plant minerals needed for a healthy life, so today humans need to add these plant minerals to the diet. The reason these minerals are so essential, they claim, is that humans cannot effectively use any nutrients without mineral cofactors. Also they point out that most centenarians live in cultures in remote and poor environments, where they live on CR diets imposed by their environments of only about 1200–1500 calories a day. Additionally, these centenarians consume large qualities of antioxidant food and drink, which includes coffee, green tea, chocolate, wines, fruit, sweet potatoes, kelp, and vegetables. Also they consume multiple types of minerals from various sources, including flood silt, dust compost, wood ash, kelp, ashes from cooking, and from the fertilizer created from wood, peat, kelp, rice straw, and ash.[2]

Thus Wallach and Lan propose that humans today who want to extend their life to at least 100 years or more should adopt this CR diet, along with a nutritional supplement that provides the essential minerals otherwise missing in the modern diet. They recommend consuming plant minerals, supplements, wood ash, and large quantities of high-grade antioxidants each day. Plus you should follow this diet without malnutrition. In other words, eat less but eat well.

To back up their approach, they provide information on the aging process that they uncovered in the course of their extensive research. For example, they describe how the life span of dogs, rats, and other species increased as a result of adding vitamins, minerals, rare earths, essential fatty acids, amino acids, and antioxidants to a commercially prepared diet of dog food.[3]

Besides embracing this CR diet, their other recommendations for health and longevity include some of the standard advice of other longevity advocates, such as refrain from smoking, participating in destructive behavior, and eating fried foods, and providing for your spiritual self. But most of all they emphasize that you have to embrace a CR diet. All of the research on increasing the life span in laboratory animals, which have been fed a perfect diet, shows that there are three known methods of increasing a maximum life span—having a lower metabolic rate through hypothermia, having enhanced antioxidant protection through genetic engineering and supplements, and having a calorie-restricted intake without malnutrition. So if humans want to achieve their genetic potential, they have to add their food through nutritional supplements and antioxidants to gain the nutritional value lacking in modern American diets and consume a CR diet.[4]

To further support this approach, Wallach and Lan provide examples of some individuals who have lived to 100, and they point out the positive results of repeated experiments of the CR diet without malnutrition on a variety of vertebrate and invertebrate species. For example, they report that diet-restricted mammals retained their youthful characteristics, including their immune function and vitality, and they showed other positive signs of health compared to their littermates who were fed a regular diet. Among other positive signs of health, they had lower circulating glucose and insulin levels, a lower percentage of body fat, reduced cell loss from the aging process, less inflation or free radical injury, and a more youthful physiology.[5] The assumption is that these traits will be reflected in humans on CR diets as well.

Having made their case for why this type of diet is so important, Wallach and Lan devote the rest of the book to specific practices that individuals should follow to increase their chances for living to 100 or more. Among their major recommendations for a longer life are the following:

- To start a CR diet, reduce your total daily calorie intake by at least 5 percent based on your age, sex, and whether you live a sedentary, moderately active, or active lifestyle; reducing your intake by 10–20 percent is ideal. Within a week you should notice that your body is more efficient in producing energy from fat, and your stamina is increased as well.

- Take steps to reduce or remove the oxygen free radicals and other reactive oxygen species (also referred to as ROS) as these are important factors in many disease processes and aging. That's because these highly radical substances set off abnormal reactions in the cells that are irreversible and energetically wasteful. You should eat fewer foods with free radicals, which are present in fried foods, trans fats, alcohol, sugar, and others.

- Significantly increase the antioxidants in your diet up to 20,000–25,000 ORAC points daily, based on the Oxygen Radical Absorbance Capacity scale that measures the antioxidant capacities of various foods. This scale was developed by scientists at the National Institute on Aging (a division of the National Institutes of Health in Bethesda, Maryland) and Tuffs University in

Boston. Some examples of foods and nutrients high in antioxidants include vitamins A, C, and E; green tea; coffee; red wine; dark berries; exotic fruits; resveratrol in red wine and red grape juice; green leafy vegetables; tomatoes; and dark chocolate.

- Avoid malnutrition, which doesn't just mean getting sufficient calories to survive. Rather you should obtain at least the optimal levels of every known essential nutrient, which is much higher than the recommended daily intake (RDI) level. Ideally your diet should include the 90 essential nutrients, including 60 minerals, that humans should have in their diet.

- Obtain many essential nutrients from supplements since humans can't manufacture certain nutrients needed for health, reproduction, and living as long as possible.[6]

So there you have it—the basic requirements for the kind of diet to adopt if you hope to live to at least 100. The rest of the book goes on to list specifically the types of foods that you should avoid or eat as little as possible, such as carbohydrates and sugars. It also lists the essential amino acids, vitamins, and minerals to include in your diet, along with proteins, which are the building blocks of cells.[7]

There is far too much detail to describe here, though you can see the specifics of the 90 nutriments Wallach and Lan recommend, along with the way to get this CR diet without malnutrition, which includes a combination of recommended foods and nutritional supplements. Plus Dr. Wallach has even more books on nutrition on his website at http://www.drjwallach.com. So to increase your life span, this is one way to go.

The *Life Extension Express* Approach

As part of his seven-step approach to living longer in *Life Extension Express*, David A. Kekich provides his recommendations for diet and nutritional supplements, which are much less complicated to follow than the CR diet with 90 essential nutrients. In brief, Kekich recommends either the Paleo or Mediterranean diet, which both emphasize proteins, leafy green vegetables, and fruits high in antioxidants, along with the most important supplements.

In Kekich's view, having a good diet is critical as the most powerful weapon for overcoming the ravages of aging. Moreover, Kekich harks back to Paleolithic times to suggest that the hunter-gatherers ate a much better diet than we do today. Not only was there a good balance between oxidants and antioxidants, but the hunter-gathers had a diet that consisted of about 70 percent raw fruits and vegetables. Unfortunately, about

10,000 years ago, with the development of agriculture, once people started growing their food, they stopped eating a wide variety of raw foods. People started cooking more and began processing their food. The problem, according to Kekich, is that "cooking destroys many vital nutrients, and overcooking can create carcinogens."[8]

The result, he claims, is that people age more quickly if they eat too much cooked food, whereas people who tend to eat more raw food often become biologically and visibly younger. Accordingly, Kekich's MaxLife company recommends that people eat about 30–50 percent of their food uncooked. How? One way is by drinking vegetable juices and eating a salad for lunch.

Yet as much as he values raw foods, Kekich recognizes some exceptions, such as the many people with digestive problems who can't handle raw foods or beans very well. He notes that many beans and legumes, such as raw soybeans, black-eyed peas, and peanuts, contain trypsin inhibitors, which block important digestive enzymes; and certain greens, such as raw sprouts and lettuce, could be a source of food-borne illness, so you should wash these and all raw foods well before eating them. On the other hand, whereas cooking's downside is that it destroys vitamins, it helps with absorbing certain nutrients, such as the carotenoids. That's why Kekich and most nutritionists recommend a combination of both raw and cooked foods. He especially recommends consuming superfoods, such as green smoothies, since greens are the one living thing that can turn sunshine into food. Moreover, these greens contain all the essential minerals, vitamins aside from B12, and amino acids that humans need for optimal health. And smoothies can taste good, which can alleviate the difficulty of staying on a raw foods diet, because many other foods taste so much better.[9]

By contrast, Kekich warns against eating grains, dairy products, and legumes because humans have not fully adapted to agricultural diets. Instead we still adapted to the hunter-gather diet, which humans have eaten for millions of years. The basis of this diet is eating cooked meat and some vegetables. So for better health and to reduce the effects of aging and live longer, that's what he recommends, along with being physically active, since that's part of our hunter-gatherer heritage as well.

In his view, echoing the view of many other antiaging advocates, the modern Western diet is a key factor in causing every chronic disease that affects humans today. Accordingly, it is time to cast that diet aside and go back to the beginnings of how the human species evolved in Paleolithic times. Back then, the diet consisted mainly of these key ingredients: lean meat, fish, vegetables, eggs, fruits, berries, nuts, and lots of water, fiber, and protein, so it can feel filling even though it's a low-calorie diet. In effect, Kekich is favoring a CR diet as well, though with the Paleolithic formula.

Should the Paleolithic diet seem too restrictive, an alternative is a Mediterranean diet, which also emphasizes fresh fruits, including tomatoes, and vegetables, along with lean red meats, fish, and poultry, whole grains, garlic, olive oil, and a glass of red wine each day. Moreover, the Mediterranean diet avoids any processed foods, which are a "no-no" since they are full of sugars and trans fats.[10]

The Mediterranean diet, which includes avoiding any grains, dairy, and legumes and eating as little as possible, has now turned up on supermarket shelves and is advocated by support groups around the world— nearly 50,000 on Meetup alone. Citing Dr. Loren Cordain, author of *The Paleo Diet*, Kekich states there are seven principles for this diet:

- Eat a relatively large amount of animal protein.
- Eat fewer carbs than in most modern diets; you can obtain good carbs from fruits and vegetables.
- Eat a large amount of fiber from no-starch fruits and vegetables.
- Eat only a moderate amount of fats, and obtain these from monounsaturated fats or polyunsaturated fats rather than from saturated fats and seek to have a balance between omega-3 and omega-6 fats.
- Eat foods with a high-potassium but low-sodium content, and don't put salt on your food.
- Eat a diet with a relatively low alkaline content.
- Eat foods that have a lot phytochemicals, vitamins, minerals, and antioxidants from plants.[11]

Like Wallach and Lan, Kekich recommends restricting the calories in your diet, noting the use of this technique on mice, rats, and most recently monkeys, which were fed less than 30 percent of their usual calorie count, resulting in a 15–40 percent increase in their life span. Applied to humans, these findings mean you should eat low-calorie foods that have high levels of the essential vitamins, minerals, and nutrients. As for carbohydrates, which he describes as humans' main fuel source and the most consumed nutrients around the world, these fall into three major categories: sugars, fiber carbohydrates, and nonfiber carbohydrates. In his view, you should eat carbs with fiber and stay away from the nonfiber carbs, which turn into sugar. The problem with eating a diet rich in carbohydrates is that this diet has been identified as a factor in many chronic diseases, including heart disease, cancer, and diabetes. So you should adopt a low-carbohydrate diet and supplement it with high-quality fiber—at least 30 grams, according to Kekich.[12]

Still another consideration is keeping a low glycemic index, which measures how much the foods you eat elevate your blood sugar level. The foods that elevate it the most are those high in simple sugars as well as low-fiber carbohydrates that quickly turn into sugar in your body. As it turns out, the low-glycemic-index foods are those in the Paleolithic diet, such as nuts, seeds, fruits, and nonstarchy vegetables, whereas high-glycemic-index foods correspond to those common in the modern Western diet, such as sugar, white bread, cookies, cakes, candy, soft drinks, jams, white potatoes, and white rice.

Conversely, proteins are good for you since they are essentially the "building blocks of your tissues, enzymes, DNA, hemoglobin, and antibodies," which maintain the normal structure and functioning of your body. While your body makes 12 of the 20 essential amino acids that make up these proteins, you need to get eight others from the food you eat. If you don't, you face three key problems—your body lacks sufficient nutrients to repair itself, your immune system is weakened, and your metabolism slows down.[13] The process sounds a little like what happens when a car doesn't get enough oil and water: at some point it overheats and stalls, and you have to fix the car as well as refresh its oil and water to get it going again.

Given these findings about nutrition, a number of scientists and physicians (such as Dr. Robert Atkins, famous for the *Atkins Diet*; Dr. Barry Sears, known for the *Zone Diet*; and Drs. Andrews, Balart, and Bethea, authors of *Sugar Busters*) have advocated what they call a high-protein or low-carb diet, and Kekich agrees with their approach.

As for fats, according to Kekich, the best ones are the polyunsaturated fats, which help reduce inflammation and serum cholesterol. These fats are referred to as the omega-3 fatty acids, and you'll find them in fish and fish oil—especially in krill oil, which is frequently mentioned in these antiaging books, as well as in flax oil, walnuts, and algae.

Another recommendation is to eat the fruits that have rich and vibrant colors, such as blueberries, pomegranates, strawberries, purple grapes, and tomatoes (which are really fruits, not vegetables). The natural pigments that give them their bright colors have a high level of flavonoids, which are antioxidants that have many health benefits, such as helping protect individuals from a variety of conditions, including cancer, heart disease, diabetes, stroke, and dementia.[14]

You should also eat a wide variety of fresh and frozen produce, especially the vegetables with bright colors, and ideally eat one to three servings of fresh fruit a day. But avoid white potatoes, which is like eating white bread, one of the high-carb foods to avoid. You should take steps to better digest whatever you do eat, such as by eating smaller amounts of food over four to six meals a day, eating slowly, chewing well, and relaxing while you eat. You can also add a supplement with friendly bacteria, called a probiotic, that helps your body absorb certain undigested starches, sugars, and minerals. Or you can eat cultured foods that are good sources of natural healthy bacteria, such as yogurt and sauerkraut.[15]

In sum, following all of these diet recommendations will help keep you fit and trim and avoid obesity or diabetes, which are largely due to eating an unhealthy Western diet. For more specific diet recommendations, see Kekich's *Life Extension Express* book.

But just diet alone isn't enough. Along with other antiaging advocates, Kekich recommends all kinds of nutritional supplements, because even with healthy eating, we don't get enough of these in our diet. One reason we need additional supplements is that over 600 human enzymes that use vitamins or minerals have to be in the body to result in a specific reaction. But because of this vast number, it is likely that everyone will have at least some mutations in their enzymes that limit their performance. Supplements come to the rescue to overcome the effects of these mutations. And presumably the supplements can work. For example, researchers have found over 50 genetic diseases that can be overcome by taking nutritional supplements. There are also literally thousands of research studies that support taking supplements.[16]

What supplements should you take? Some of the commonly accepted recommendations are:

- Use antioxidants. Among other things, these reduce the risk of heart disease by 26–46 percent and cut down the risks of getting cancer.[17]
- Minimally take a high-potency multivitamin tablet along with essential fatty acids each day. These essential acids are the omega-3 fatty acids, which include fish oil and krill oil. Krill oil is particularly favored by nutritionists since it has very powerful antioxidants not contained in fish oil. This fish oil should be in a capsule because fish oil is very perishable, which causes oxidation that results in free radicals circulating in your body.[18]
- Other commonly recommended supplements include vitamin D3 , resveratrol (which is found in red wine), folic acid, ginkgo biloba, and A-L carnitine; plus some or all of the six key antioxidants: vitamins C and E, coenzyme $Q_{10,}$ glutathione, lipoic acid, carnosine, and vitamins K1 and K2.[19]

It is beyond the scope of this book to detail all of the supplements and their recommended daily dosages, which can be found in Kekich's *Life*

Extension Express book and in other books on antiaging and longevity. Kekich lists about 50 different vitamins and minerals to take, and you can obtain the recommended supplements from a number of companies that sell these, including the Life Extension Foundation.

The *Methuselah Project*

Another example of the use of diet and supplements to live to 100 and beyond is offered by James Lee's *Methuselah Project*. In his book, he writes about both the "life extension" and "life optimization." He notes the same general reasons for aging noted by other antiaging advocates, such as the senescence of dividing cells, the shortening of telomeres, the accumulation of cellular damage due to free radicals, and inflammation due to trans-fatty acids, such as omega-6-rich vegetable oils and sugar.[20] Lee also agrees with the principle that restricting calories extends life, and he further points to research that suggests that a type of protein called a "sirtuin" affects several aspects of longevity, including programmed cell death, inflammation, and cellular aging. But until a commercial drug is available to target certain parts of these proteins, he recommends resveratrol, calorie restriction, or cardiovascular exercise, which helps activate these protein segments. Additionally, apart from recommending the value of exercise, reducing stress, getting socially connected, and good sleep, Lee suggests that the best option for living longer is to reduce and reverse oxidative damage by having a good diet and using nutritional supplements.[21]

One way to do this is eating those foods that have a high score on a globally standardized method for assessing a food's ability to get rid of free radicals based on its oxygen radical absorbance capacity (ORAC) score. Based on this test, the higher a food's ORAC score, the better it is in antioxidant activity. Though Lee acknowledges that the link between an ORAC score and a particular biological benefit has not been proven, many scientists believe the score is a valuable guiding principle on what to eat. In Lee's view, given that the ORAC score is a guide to antioxidant activity, and dietary antioxidants can prevent and reverse oxidative damage, it is good to eat foods with a high ORAC score. In fact, some radio commercials and Internet ads now promote oxidant nutrient supplements to combat oxidants—one more example of the way companies are increasingly trying to promote antiaging products to health-conscious consumers.

What foods score the best? Lee notes that certain herbs and spices are especially high in antioxidants, such as cloves, cinnamon, oregano, and turmeric. Their scores range from about 160,000 to 315,000, even higher than antioxidants in the foods themselves. Some of the foods with the highest scores include dark chocolate, pecans, walnuts, hazelnuts, raw cranberries, artichokes, and assorted beans—kidney, pink, and black beans.

Several vitamins and supplements on Lee's list of antioxidants include vitamins C and E and krill oil—which comes largely from wild-caught salmon. Like other antiaging advocates, Lee is a strong supporter of the polyphenols and flavonoids, which come especially from a diet that includes a lot of multicolored fruits and vegetables as well as from tea and berries. In fact, he suggests that it is best to get antioxidants directly from real foods rather than from supplements, and he especially favors berries because they are relatively low in fructose, whereas many other fruits are high in it, which could outweigh the antioxidant benefits if you consume too much fruit.[22]

Still other recommended food and nutritional supplements include resveratrol, which comes from wine, though drinking more than two glasses of wine per day can turn into a negative for your health due to the effects of alcohol. Lee also emphasizes the importance of glutathione to reduce oxidative stress and detoxify, and he points out that glutathione helps make other antioxidants, such as vitamins C and E and lipoic acid, even more effective. The best way to get glutathione is to use supplements in addition to getting it from cruciferous vegetables such as broccoli and sulfur-rich foods such as garlic and onions. Some suggested supplements are N-acetylcysteine (also known as NAC), alpha lipoic acid, and selenium, a trace element that assists glutathione in neutralizing free radicals.[23]

Still another approach to antiaging is to take supplements to reduce the levels of inflammation in the body. Examples include omega-3 fatty acids and curcumin. Omega-3, which comes especially from fish oil, krill oil tablets, and healthy seafood, is very helpful in contributing to the health of myelin, the fatty sheath that covers the nerves in the brain. Omega-3 balances out the harmful effects of having too much omega-6, which comes from grain-based products.[24]

As for curcumin, which comes from turmeric, it is not only an anti-inflammatory and antioxidant but also helps prevent liver and kidney toxicity. It also can be used to treat numerous diseases including diabetes, multiple sclerosis, Alzheimer's, septic shock, cardiovascular disease, lung fibrosis, and arthritis, and it has anticancer benefits. Its anti-inflammatory properties are especially important: it reduces the potential for heart disease since inflammation of the cardiovascular system causes the majority of incidences of heart disease.

Finally, Lee recommends vitamin D as another important anti-inflammatory, which also inhibits telomere shortening due to reduced inflammation. It also inhibits certain types of cell proliferation, which is why it is used in many antipsoriasis creams, as cell proliferation is associated with that skin condition. Besides obtaining vitamin D through a supplement, Lee recommends getting at least 20 minutes of direct sunlight exposure daily.[25]

Like many other antiaging advocates, Lee recommends minimizing the consumption of sugar, especially fructose, since they both promote inflammation and the production of advanced glycation end products (AGEs). These are usually created when glucose binds with a protein, resulting in cellular damage. Since sugar is addictive, you can experience withdrawal symptoms if you quickly cut down on your sugar intake, like someone who suddenly stops taking a drug like heroin. So you should gradually change your diet from foods high in sugars to foods with more neutral or salty flavors. Another strategy to reduce AGEs is to reduce the amount of packaged and processed foods you eat because they tend to have a high level of sugar.[26]

Lee also recommends taking steps to maintain a healthy liver, also central to longevity, along with maintaining a healthy heart and brain. The most important supplements include N-acetylcysteine (NAC), curcumin, and alpha lipoic acid (ALA); plus milk thistle (also known as silymarin), an herb that has been used for over 2,000 years as a natural treatment for assorted diseases (including kidney and gall bladder problems and cancer), and it lowers cholesterol levels. It comes from the seeds of the milk thistle plant, which have been used to create a variety of medicinal extracts, powders, and tinctures.[27]

Lee concludes his book with a list of what to include in a healthy diet. Besides his already mentioned suggestions, he recommends taking some supplements and drugs, called nootropics, to keep the brain young. These nootropics enhance learning and memory, protect the brain from damage, improve brain functioning, and are relatively safe without serious side effects. The major supplements and drugs include acetylcholine, the brain's most common neurotransmitter, dopamine, another neurotransmitter, choline, NAC, B-group vitamins, and curcumin.[28] For more specifics, refer to his book.

In short, as reflected in Lee's book and in the recommendations of the other antiaging advocates cited, there is general agreement on what is required to have a healthy diet and how to supplement it with various vitamins, minerals, and chemical extracts. Doing so may seem daunting, because there are hundreds of these items. But many longevity advocates recommend or have their own companies, which provide a variety of antiaging supplements in the form of pills, capsules, creams, and other products to help you live longer. Then you need to add in the other methods for living longer, which I'll describe next.

Weight Loss and Calorie Reduction

Along with eating certain types of foods and taking nutritional supplements, many antiaging advocates advise keeping one's weight down. Aside from the CR diet, already discussed, they recommend a general reduction in calories, such as by eating smaller portions. They also recommend avoiding sugar and following a low-carb diet rich in vegetables and fruits to help keep down one's weight. Additionally, the exercise and fitness activities discussed in chapter 6 are another way to achieve weight loss.

As part of their Nine Steps to Living Well Forever program, Ray Kurzweil and Terry Grossman include calorie reduction and weight loss as part of the plan. In turn, this goal of losing weight and staying thin is deeply etched into the national psyche, although two-thirds of the adult population is deemed overweight and one-third is obese.[29] But many are trying and often failing to lose weight each year. That's why about 45 million Americans are on a diet each year, according to the Boston Medical Center.[30] No wonder TV shows about losing weight, such as *The Biggest Loser*, get top ratings and all kind of diet products line the supermarket shelves. A search of Google reveals dozens of diet plans, including the Nutrisystem Weight Loss program, the Atkins Diet, the 14-Day Ketogenic Diet Plan, the ,1200 Calorie Diet Menu, and gluten-free and diabetic diet plans from EatingWell.

The December 2016 issue of *Redbook* even featured an article on "33 Top Diet Plans That Are Actually Worth Trying."[31] These include:

- the Wild Diet, where you eliminate all processed foods
- the Taco Diet, where you can put anything in a taco for 30 days
- the Disassociated Diet, which dates back to 1911, where you don't combine acidic foods (such as meats, fish, and dairy) with alkaline ones (such as legumes, vegetables, and nuts), which is supposed to be easier on your digestive system
- the Weight Watchers Diet, in which you follow a daily allotment of foods based on calories, saturated fats, protein, and sugar, while you can eat fruits and veggies freely
- the Mediterranean Diet, which is based on eating fruits, vegetables, whole grains, nuts, fish, olive oil, and wine
- the DASH Diet (which stands for Dietary Approaches to Stop Hypertension), which emphasizes fruits and vegetables and reduces sodium, fat, and saturated fats

- the MIND Diet, which combines the Mediterranean and DASH diets, which not only offers weight loss but also improves your brain, which is designed to help prevent Alzheimer's disease by emphasizing leafy green vegetables, whole grains, olive oil, and wine
- the TLC Diet (which stands for Therapeutic Lifestyle Changes), which is low in saturated fats and cholesterol and designed to reduce cholesterol and the risk of heart disease
- the Paleo Diet, which is big on the proteins from meat our Paleolithic ancestors used to eat
- the Vegan Diet and Vegetarian Diet, which eschews most or all animal products
- the Zone Diet, which is a low-carb plan, where you only eat low-fat protein, nonstarchy vegetables, small amounts of fruit, and a small amount of healthy fat, like olive oil

And that's just the first 12 diets. There are even special diets such as the Fertility Diet for expectant mothers; the Nutrisystem Diet, promoted by spokesperson and entertainer Marie Osmond; the Biggest Loser Diet, based on the show of that name, which features fruits, vegetables, and lean protein; the High Blood Pressure Diet; the Anti-Inflammatory Diet; and the Soup Diet. It's as if you can pull together any combination of recommended foods, perhaps add in some nutritional supplements, and you have a new unique diet plan. What makes most of them work is they follow the guidelines for eating healthy foods and getting the additional supplements for the minerals, vitamins, and other chemicals you don't get from food alone. Voilà, you have a new diet, and perhaps a line of diet products for sale, for about two-thirds of the population who are overweight or obese.

Alternatively, you might do fine dieting on your own if you just select from those foods recommended by the antiaging advocates previously described. Add in some supplements to get the needed minerals, vitamins, and chemicals, and keep your own weight down by not eating too much of a good thing.

Why Losing Weight Is So Hard

Given the fact that so many people are overweight and constantly trying to lose weight, the question is "WHY?" One reason is that people simply eat too much. For example, Kurzweil and Grossman point out that as of 2004, Americans consumed an average of 2,750 calories a day, 500 more calories per day than they ate in 1970, resulting in an increase of about 25 pounds for both men and women.

The other reason for being overweight and not being able to take off the weight is that most people go on a diet to temporarily change their eating habits, whereby they temporarily deprive themselves of the foods they really like. But once they lose the desired weight, they go back to the way they used to eat, so they soon put the pounds back on and sometimes even more. As a result, two-thirds of all dieters gain back any weight they lost within a year, and 97 percent gain it back with five years. In fact this kind of up-and-down dieting, called "yo-yo dieting," may be more damaging to your health than if you maintain the same number of excess pounds during this time. So the real solution is to change your way of eating for the rest of your life.[32]

Just going on a diet for a brief period or switching from diet to diet in the hope that something works isn't likely to work at all. Rather you need to reduce the number of calories you eat in order to lose weight gradually without feeling hungry or deprived. Then you need to continue to keep that weight off. This way you end up with a leaner, healthier body for the rest of your life.[33]

Changing the Way You Eat Permanently

A first step proposed by Kurzweil and Grossman is to get to your goal weight and stay there permanently by consuming as many calories a day as you would if you were already at this goal. Then stick to this calorie level for your target weight, and you will gradually get to your goal. At the same time follow a low-calorie diet with better food choices: fewer sugar and high-fat foods and more low-calorie foods, such as leafy green vegetables and certain fruits like berries.

What is your ideal weight? Kurzweil and Grossman provide a chart based on your height and whether you have a small, medium, or large frame, which you can determine by measuring your wrist size. To find your maintenance calorie level, first determine your activity level, based on whether you are sedentary, moderately active, or very active. After that, look for your target weight on a chart, then find the number of calories to maintain that weight based your activity level.[34]

Losing weight, though, shouldn't be just about weight loss, and you shouldn't go to extremes in reducing your daily calories. As a rule, you shouldn't eat less than 13 calories a day for each pound of your optimal weight, so women should eat at least 1,000 calories a day and men at least 1,200 calories a day. Your goal should be good health, not just losing weight; if you make sustainable lifestyle changes that result in better overall health, you will automatically lose weight. Moreover, when you lose weight gradually, you put less stress on your body's organs and systems than if you rapidly lose weight.[35]

By the same token, to make any weight-loss plan work in the long term, choose foods you can continue to eat once you achieve the desired weight.

Finally, Kurzweil and Gross recommend combining this weight-loss plan with other principles of healthy eating, such as eliminating sugar and refined carbohydrates and exercising, as being sedentary contributes to weight gain. They also recommend going beyond your goal weight by engaging in a program of calorie reduction by eating less. Like other anti-aging advocates, they describe the many studies where different animal species have lived longer due to CR diets—more than 2,000 studies at the time they were writing in 2009. They recommend a 10–20 percent reduction in calories, which is relatively safe if you obtain an optimal amount of nutrition by eating foods high in fiber and nutrients but low in calories, such as low-starch vegetables. Also, avoid high-starch foods, maintain a healthy balance of carbohydrates, lean proteins, and good fats, and take the appropriate nutritional supplements.[36]

There is even an organization devoted to calorie restriction, CR Society International (www.crsociety.org), which has an Interactive Diet Planner, originally developed by the late Dr. Walford. It contains a chart so you both restrict your calories and get the necessary nutrients. As a guide for eating right, the diet planner database lists the nutrient values for 3,000 foods, drawn mostly from the listings of the US Department of Agriculture and from German and Japanese databases. You can use the plan to create a food list and see the nutritional totals so that you can determine what foods to drop or add in order to get the necessary nutrients for a certain number of calories, based on your calorie goal. You can also add foods to the database or analyze foods for their nutrient value, based on 6,000 foods from the USDA database.[37]

The society was originally founded as a nonprofit in 1994 by Brian M. Delaney, Roy Walford, and Lisa Walford, and it is based on the work of several thousand individuals who practice, support, and conduct research on calorie restriction to slow down the aging process.

CHAPTER SEVEN

Living Right to Live Longer

Besides getting the right nutrition and keeping your weight and calories down, another strategy is to live a healthy lifestyle, which consists of five main components. You need to:

- regularly exercise and stay fit
- avoid risky behaviors, from abusing drugs and alcohol to engaging in dangerous activities
- manage your stress so that you stay calm and relaxed
- have a good, positive attitude
- reduce as much as possible your exposure to toxins in the environment or take steps to reduce these toxins in your system

Some antiaging medicine can help too, and I'll discuss that in a separate section.

Exercise and Fitness

Exercising and staying fit is a key principle of living longer, widely advocated by the antiaging writers and researchers.

For example, in *The Methuselah Project*, James Lee briefly digresses from recommending various food and nutritional supplements to extolling exercise, particularly for its benefits for the brain. That's because you experience a series of biochemical reactions so that you feel a reward of pleasure when you exercise.

In particular, exercise stimulates a certain protein, BDNF (brain-derived neurotrophic factor), which helps your brain cells thrive, and it helps trigger the birth of new neurons, called "neurogenesis." In addition, exercise stimulates the secretion of BDNF in the hippocampus, which contributes to better memory recall. If you exercise, even for a short time, such as running for 30 minutes a day, that will increase the production of BDNF in your brain, enabling you to think more effectively and improve your mood.[1]

Still another benefit of exercise is that it increases the levels of monoamines—serotonin, dopamine, and norepinephrine—that help elevate your mood so you feel happier. Exercise also contributes to an improved blood flow to the brain, which increases the delivery of oxygen. Exercise also improves sleep, which is when the majority of repair in the brain occurs, especially when you are most deeply asleep, during slow wave (stages 3 and 4 NREM) sleep. As a result of this increase in slow wave sleep, you feel more refreshed when you wake up, and your brain can make any repairs more quickly. Then, too, you have a better sleep, since exercise burns away a lot of the hormones and neurotransmitters (such as cortisol and norepinephrine), which are associated with stress.

What kind of exercise should you do? Lee suggests doing a variety of cardiovascular exercises, strength training, and stretches. Even just walking can help, especially to overcome symptoms of depression. But don't overdo it, since he reports that researchers have found that the best kind of exercise has a short duration and is of a high intensity.[2]

You'll find extensive examples of different types of exercisers in David A. Kekich's book, *Life Extension Express*. He warns that if you aren't active enough, you may experience "bone loss, poor cardiovascular tone, decreased telomere length, and increased incidence of heart disease, cancer and diabetes"[3] compared to people who are physically active. Starting an exercise program will help at any age, as documented in a number of research studies. But you have to continue doing it for an extended period of time, ideally for at least five years, as researchers found in a study of midlife men in Sweden. If you exercise enough for a long-enough time, exercise contributes to a longer life. A major benefit is that you cut down your heart attack risk, and researchers have found that regular exercise helps "maintain the levels of hormones that typically decline with age."[4]

The two main types of exercise include strength (or anaerobic) training and cardio (or aerobic) training. One irony of this emphasis on exercise as one of the keys to living longer is that the gyms, fitness clubs, and other organized programs that offer different types of exercise don't normally mention living longer as one of the benefits. Rather, the appeal is based on being more fit and healthy, losing weight, strengthening the abs and other muscles, having more endurance, being more attractive, and meeting other people who want to be more fit and healthy, as emphasized at dozens of presentations I have attended by fitness coaches and sports trainers at networking events and gym tours. Yet, for those interested in living longer, exercise is one of the major techniques to increase one's life span in a healthy way, by using both strength and cardio training as well as exercises to maximize your brainpower. And if you enjoy competing, there are local and national competitions to show off your abilities in each of these areas.

Strength Training

Strength training, also referred to as "resistance training," is commonly developed by weightlifting. The idea is that you strengthen your muscles by adding more and more weights to build up your resistance power. Among other things, this training is good for overcoming oxidative stress, due to the oxidizing free radicals, and it contributes to creating greater bone density and strengthening the tendons, ligaments, and joints. It also helps prevent osteoporosis—a condition where the bones become fragile and brittle from loss of tissue, and it helps prevent connective tissue damage. These effects are further enhanced by taking hormone supplements, which are especially helpful for women, since their natural hormones decrease at a faster rate. Strength training also benefits you by:

- reducing fat and making you more flexible
- reversing the aging effects on the skeletal muscles so that you feel more youthful
- increasing your muscle mass so that you can engage in everyday activities more easily
- strengthening your ligaments and tendons, which will protect your joints from arthritis
- increasing your levels of the anabolic hormones, such as testosterone and DHEA (dehydroepiandrosterone), a hormone from the adrenal gland and brain, which leads to the production of androgens and estrogens, the male and female sex hormones[5]

Unfortunately, as Kurzweil and Grossman point out, even fewer people engage in regular strength training than in aerobics. Individuals over 65, who most need to maintain their muscle mass, are even less likely to engage in this training. (Only about 12 percent of this older population does so.) This kind of training is important, because people who don't do strength exercises lose up to 40–80 percent of their muscle mass between

20 and 80 years of age. But strength training helps counteract the natural tendency of muscles to shrink as one ages, and it will increase the level of some of your hormones.[6]

To get all these benefits, as sports and fitness coaches and trainers commonly advise, you should gradually work up to a good fitness level by working out on a regular basis, perhaps two or three times a week. Ideally, work on the muscles in different parts of your body so that you can give the muscles you recently exercised a rest. Include stretch exercises at the end of your session for flexibility, and include a variety of exercises, such as yoga and Pilates. More specifically, some of the basic guidelines for strength training, according to Kekich, include:

- Do exercises for all major muscle groups at least twice a week, but don't exercise the same group two days in a row.
- Start with a low amount of weight or no weight and gradually increase the amount of weight.
- To progress, increase the number of times you do an exercise, and at a later session, increase the weight.
- Keep your back and shoulders straight when you bend forward so that you bend from your hips and not your waist.[7]

Cardio Training

Cardio (or aerobic training) includes a variety of exercises, such as running, jogging, bike riding, walking, swimming, and using treadmills and stair-stepping machines. These exercises are designed to raise one's heart rate and thereby increase the health of one's cardiovascular system. The term "aerobic" was originally coined by Dr. Ken Cooper in 1968 to refer to exercising done in the presence of oxygen.

The purpose of these exercises is to participate in a physical activity that gets the heart and lungs working more intensely to increase the flow of oxygen to the body. The result is improved cardiovascular fitness and health, while the workout benefits almost every tissue in the body (e.g., reducing abdominal fat, increasing the natural supply of endorphins, and helping relieve mild depression).[8]

Preferably use these exercises to increase your heart rate for at least 20 minutes a day for the best effect. The benefits, which also contribute to an increased life span, include more energy, improved breathing, more energy, better heart health, lower blood pressure, decreased cholesterol, less stress, better sleep, improved mood, and increased mental functioning.[9]

In fact, it's good to combine weight training with some aerobic activities, since weight training alone might increase one's chances of a heart attack or stroke. Ideally, you should engage in a combination of these exercises 6–7 days a week for 20–45 minutes per session. If you do extensive training one day, then take a day or two off to rest. And you needn't engage in these exercises on your own, since a whole industry of health clubs, gyms, and trainers has evolved to help you through the process in a fun, supportive way, and you can join a community of others who are similarly committed to regular exercise. Groups devoted to various types of exercise activities commonly don't stress the antiaging benefits; rather, they emphasize fitness while having fun.

Getting Started and Measuring Your Results

Whether you join an organized health and fitness group or exercise on your own, a key to doing so successfully is to gradually build up your endurance, especially if you have been inactive, starting with as little as five minutes of activity a day. Then, gradually build up your strength or stamina. You might also benefit from interval training, in which you engage in a series of low- to high-intensity exercise workouts combined with occasional rest or relief periods of lower-intensity activities or rest.

To chart your progress, measure how well you are doing. Kekich recommends two methods: the target heart rate scale (THR) and the Borg Scale, which is a subjective rating.[10] Another type of measurement described by Kurzweil and Grossman is determining your maximum predicted heart rate (MPHR), which is calculated at 220 minus your age.[11] You can also access apps through your smartwatch or smartphone that measure your activity. You might also use the National Institute of Aging's chart[12] that indicates the desired range for heart rate during endurance exercise based on the number of beats per minute for different age groups.[13]

How often should you engage in these activities? According to the Centers for Disease Control and Prevention (CDC) and the American College of Sports Medicine (ACSM), all adults should participate in these aerobic activities for at least 30 minutes at least three times a week (ideally every day), and preferably you shouldn't have more than three days between sessions.[14] You should structure your session into four phases:

- Warm up for 3–5 minutes to walk around and warm up your muscles to increase the blood flow to them plus some stretching.
- Engage in the exercise to get your heart rate to 65–85 percent of your MPHR. Your session should last for at least 20 minutes and ideally 30–40 minutes,

since the benefits to your heart begin after 20 minutes of continuous exercise.

- Engage in a cooldown, walking slowly for about 3–5 minutes so that your heart rate can return to normal.
- Participate in stretching, since your tendons, ligaments, and muscles tend to tighten up after an exercise session.

Initially begin this exercise session slowly, build up gradually, and make these exercises a habit. You can find specific techniques in many exercise books and videos (search on Google or YouTube), and Kurzweil and Grossman have a section on different types of exercises, including stretches for more flexibility. They conclude with a suggested basic exercise program that combines aerobics, strength training, and flexibility.[15]

Improving Your Brain Functioning

All of these exercises for strength and improving circulation will improve your brain function as well. Among other things they will strengthen the neurons and improve the flow of blood. oxygen, and nutrients to your brain. But you should exercise your brain too, since your brain can atrophy if it isn't used. For instance, keep learning something new, participate in problem solving, read, and play games such as chess that challenge your intellect. The results will include increased vitality, alertness, and thinking ability, and better retention; and you will be able to think more clearly.[16]

The Importance of Regular Exercise and Overcoming Our Resistance to It

In *Transcend: Nine Steps to Living Well Forever*, Ray Kurzweil and Terry Grossman similarly advocate exercise as one of the nine steps, and they emphasize that it will not only make you stronger but help prevent disease so that you can live forever. They provide some examples of research showing the value of regular exercise. For example, in another 2007 study, Italian researchers showed that the more exercise you do, the less risk you have of developing Alzheimer's disease. A 2008 study of 12,000 Danish adults found that combining regular physical activity with moderate alcohol consumption reduced the risk of heart attack by 50 percent and decreased the mortality rate for all causes by one-third.[17]

As Kurzweil and Grossman point out, the human body is designed for regular exercise. In prehistoric times, life in the Stone Age required vigorous physical activity, including walking and running long distances, climbing, and throwing to procure food or avoid danger.[18] So the need to exercise today can be traced back to the beginnings of human history. Genetically we are still programmed to need both regular exercise and rest.

Kurtzweil and Grossman also point out that most people don't like to exercise, a predisposition that goes back to prehistoric days, when humans' desire to rest whenever they weren't driven to find food or escape danger conserved calories, which in turn helped their often marginal food supply last longer and thereby contributed to their survival.[19] The way to counter this predisposition is to make exercise sessions more enjoyable and pleasurable. One motivator is that people soon find that they "begin to look and feel better." Another motivator is that exercise increases the levels of endorphins, which are chemicals in your body that are associated with pleasure. If you combine exercise with activities you enjoy, such as walking along the beach or exploring a park with a friend, that will help you want to exercise.

Thus even if you don't like to exercise, do it anyway, and find ways to make it enjoyable so that you will stick with it. Start slowly but keep exercising consistently, and that will contribute to both health and longer life.

Living a Healthier Lifestyle

Your lifestyle affects how long you live, so living a healthier lifestyle will help extend your life span. For example, if you engage in a lot of risky behaviors, such as participating in extreme sports, driving fast cars, abusing drugs and alcohol, engaging in dangerous criminal activity, and otherwise living on the edge, you are more likely to slip over and end up very dead, so there is no coming back.

Conversely, if you make changes in your life to live a healthier lifestyle, which includes having a better diet, keeping your weight down, and exercising, that can extend your life. For example, in the Interheart study, about a dozen researchers studied 30,000 subjects in 152 countries and found that changing one's lifestyle could prevent about 90 percent of all heart disease, which is responsible for more premature deaths than any other illness. In turn, the same lifestyle changes that prevent or reverse heart disease can help prevent or reverse many other chronic diseases.[20]

Some key lifestyle factors that can shape how long you live include:.

- Avoid accidents while driving. You may not be able to avoid every crazy, drunk, and otherwise incapacitated drivers, but you can reduce your risk if you decide not to drive while incapacitated by drugs, alcohol, or a lack of sleep. Also, keep up your car's maintenance, since faulty brakes, tires, and other car problems can lead to accidents or stalls that leave you vulnerable while stopped on the side of the road.

• Reduce the risk of household accidents, since these are a major cause of accidental deaths, such as from slips and fires. One way to check for risks is to walk through your house, look for problems, and correct them as soon as possible. Other strategies include having a smoke detector in every room, and only use products in a well-ventilated area. Also fix any unsafe electrical connections and install an effective home security system.

• Avoid initiating or engaging in confrontations that can escalate and result in physical attacks. For example, if you experience road rage from another driver, it is often best to back away rather than risk being attacked or killed by an angry driver.

• Stop smoking. Smoking is one of the major preventable underlying causes of death throughout the world. Among other harms, it is a key cause of cancer—40 percent of all cancers, according to researchers. Moreover, smoking breaks down collagen in the skin, which leads to premature wrinkling.[21]

The data from the U.S. Centers for Disease Control and Prevention is even more disheartening. It shows that male smokers lived 13.2 years less and female smokers lived 14.5 years less than their nonsmoker counterparts and that the overall mortality among both male and female smokers in the United States is about three times higher than that among those who never smoked.[22]

Also be careful about your exposure to the sun. This exposure can be very healthy if you get the right amount, but it is not healthy to get too much or too little. Exposure to the sun between 10 a.m. to 2 p.m.) is associated with higher levels of damaging UVB radiation; long-wave ultraviolet (or UVA) radiation is stronger at other times (associated with melanoma or skin cancer). If you do spend more than 20 minutes in the sun, use a sunscreen or protective clothing.[23]

Getting a good night's sleep is also important for longevity, since sleep deprivation has negative effects on your body. Among other things, not getting enough sleep for even one night can trigger the cells to produce inflammation that damages tissues. This inflammation occurs, according to a UCLA Cousins Center research project, because your body experiences sleep deprivation as stress and responds by manufacturing deadly stress chemicals. As you sleep less for more days, you increasingly impair your ability to function, and the damage is cumulative. In addition, a lack of sleep increases mood swings, stress, irrationality, and blood pressure;

reduces your ability to adapt to change; impairs performance; saps your energy; slows reactions; and impairs memory, judgment, and decision making. It also contributes to diabetes and obesity; weakens your immune system; contributes to depression, cancer, heart disease, and stroke; and interferes with your ability to think clearly. Being tired can also be especially dangerous when you are driving or working around heavy equipment, and it has been implicated in a number of vehicle and industrial accidents.[24]

Doctors commonly recommend that the average person should get 7–8 hours of good-quality sleep each night. The recommended amount of sleep varies, since some people need more (even 9 or 10 hours) and others need less (only 5 or 6 hours) per night. You can tell that you are getting enough sleep if you feel alert and can think clearly. You aren't getting enough if you find yourself feeling tired and drifting off to sleep during the day. If you get too much sleep, you might find yourself feeling groggy and lethargic. It also helps to have a regular sleep schedule and adjust it to your more natural sleep cycle, such as whether you tend to be an early or late riser.

Taking care of your teeth can contribute to longevity too. Among other things, if you don't take care of your teeth and gums, the number of bacteria in your mouth can soar. The potential damage for your health is huge. As Kekich points out, about 75 percent of U.S. adults have some degree of periodontitis, a chronic inflammation of the gums, which predisposes people to a variety of conditions including diabetes, respiratory diseases, obesity, osteoporosis, and cardiovascular diseases, resulting in heart attack or congestive heart failure.[25]

Having good relationships also contributes to longevity because of the psychological well-being you feel by having these relationships. By contrast, being lonely can reduce your life span. Then, too, healthy positive thinking and self-esteem contributes to longevity as does being around people who are positive, upbeat, and health conscious, which will help you maintain that attitude. By contrast, if you are around people who pursue an unhealthy lifestyle, such as by smoking, eating and drinking too much, or having a negative outlook on life, you will reduce your life span.

Having enough money is still another longevity factor, since it enables you to pay for treatments that can extend your life, such as future biotech and high-tech methods that increase longevity and reverse aging. Start saving or investing wisely to raise the money you might need to pay for the treatments to extend your life.

Avoiding Risky Behavior and Places

Reducing the risk of losing your life under various potentially dangerous situations is another strategy for living longer:

- Avoid dangerous neighborhoods and situations. For example, if your area is
  hit by a major hurricane, flood, earthquake, fire, or other danger from
  nature, evacuate when ordered to do so. Likewise, to protect yourself from
  social unrest, acts of terrorism, or other social conflicts, prepare in advance
  by packing a survival kit to have on hand if needed, which should contain
  14–30 days' supply of food and water; candles, flashlights, and batteries;
  and first aid supplies.[26] Install a generator and water filter in your house in
  case the power system goes out. If you can, choose to live in a generally safe
  neighborhood.

- Be very careful if you have to go to a tough neighborhood. Stay safe by traveling in well-lit areas and where there are other people or avoid going to
  such neighborhoods entirely. Carry a phone in case of emergency.

- Reduce or avoid risky behaviors that could shorten your life, or at least
  assess the odds and decide what you want to risk doing. For example, if you
  really like participating in extreme sports, recognize the risks you are taking and be happy about the odds you are facing. Also be aware that the risks
  you take are cumulative and additive; the more you do something risky, the
  more you increase your chances that your chosen activity might be fatal.[27]

- Take some time to assess your life for what you are doing that is contributing
  to health and longevity and what might be detracting from that. Then think
  about what you can do differently going forward to help extend your life.

Avoiding Toxins in Your Environment

Another factor that can shorten your life is the chemicals and toxins you
are exposed to in your environment and in the products and services you
use. It is not possible to avoid all of them, since these chemicals and toxins
are all around us, but you can take steps to at least reduce your exposure.

Among these toxins are the many toxic metals that are found throughout the environment and invade your body. Among them are mercury,
aluminum, cadmium, arsenic, lead, and nickel. Even though many environmental dangers exist, you can detoxify your system in various ways.

One way is eating a healthy diet and getting regular exercise, as previously described. Another recommendation is to drink clean water to stay
hydrated, ideally by drinking eight or more glasses of water a day. Otherwise, being slightly dehydrated can interfere with the functions of critical
cells. Ideally, don't drink regular tap water or drink soft drinks or energy

drinks in place of water, since the tap water can contain chemicals. It can also carry microscopic bacteria, viruses, or parasites and can spread more serious diseases like hepatitis and cholera. Moreover, chemicals used at water-processing facilities during the treatment process (such as chlorine to kill bacteria, viruses, and parasites) and long-term exposure to these chemicals has been linked to various medical problems, such as cancer and problems with the liver, kidneys, and reproductive tract.[28] Rather, it is best to use a water-filtration system, and you might additionally use a water ionizer to alkalize the water, as recommended by some experts.[29] Check online or your local hardware store to decide on the best device for you.

You have to be careful about soft drinks and energy drinks as well, since they can contain sugar, caffeine, or other unhealthy additives.

You can also reduce your exposure to a number of airborne toxins, according to Kurzweil and Grossman in *Transcend: Nine Steps to Living Well Forever*. Obviously we have to breathe air in and out every day, but we can limit our exposure to these toxins. Some of these strategies include[30]:

- Don't use conventional detergents and oven, toilet, and glass cleaners, since they contain toxic ingredients. Instead use nontoxic substances such as lemon juice, baking soda, borax, and vinegar or use commercially available products for environmentally friendly companies, such as Earthy Friendly Products (http://www.earth-friendly.com), Ecover (http://us.ecover.com), and Seventh Generation (https://www.seventhgeneration.com).
- Use botanically based pest control products to combat insects rather than products like Raid that most supermarkets stock.
- Use stand-alone air filters to remove both large visible particles and microscopic particles, such as bacteria, viruses, and fungi.
- Have lots of plants in your house, since they help reduce airborne toxins that air filters can't eliminate, such as benzene and formaldehyde.
- Don't allow smoking in your house.
- Be careful about any hazards you might face in your home and at work and seek to reduce your exposure. For example, workers in dangerous jobs (such as logging, construction, and agriculture) should understand and follow the most current safety procedures. As for home hazards, be careful in handling solvents and pesticides by wearing the appropriate safety gear (such as coveralls, rubber gloves, and a face mask) and dispose of hazardous cleaning supplies through your city recycling program.[31]
- Limit your exposure to electromagnetic radiation. Some of this radiation comes from satellite transmissions, communications towers, cell phones, and various electronic devices from TVs and computer monitors to hair dryers and electric razors.[32] While you can't completely avoid these devices,

you can reduce your exposure. Some ways to do this, according to Kurzweil and Grossman, include:

- using a Bluetooth earpiece with your cell phone, since that has less radiation than your cell phone
- spending as little time as possible with high-power electrical devices, such as hair dryers and electric shavers
- sitting at least 10 feet away from television sets[33]

And if all of these admonitions about how to avoid toxins, radiation, and chemicals in our everyday environment get you down, that's another source of decreasing your longevity, because stress and anxiety can take away years from your life. By contrast, if you relax and have a good attitude, that will add to it.

Your Attitude, Relaxation, and Overcoming Stress and Anxiety

A number of psychological factors also contribute to your living a long life or deducting years from your life. The major factors include having a good attitude, learning to relax, managing your stress, and overcoming feelings of anxiety. These factors are interrelated and derive from having a positive, optimistic outlook on life, where you tend to be calm and placid and don't become overly stressed, angry, or anxious. You might feel some stress when you try new challenges and get out of your comfort zone, such as when you start on a new project or job or have to speak in public. Likewise you might feel some anxiety and nervousness about beginning a new venture. But that is good stress that helps you overcome obstacles and achieve success in life, and such anxiety is natural. In fact, this anxiety might help you perform even better since it gets your adrenalin flowing. By contrast, the other kind of stress is when you are upset, feel helpless, fearful, or experience angry, vengeful feelings—and these are the kind of stress that can take years from your life.

What can you do to nurture a positive attitude and learn to relax while overcoming harmful stress and anxiety? I'll discuss some of the basic methods here; beyond that there are all kinds of psychological techniques, programs, support groups, and other methods that have fueled thousands of books, courses, and workshops on how to live a happier, more fulfilling life and overcome whatever is causing your stress.

Having a Positive Attitude

The importance of having a positive, upbeat attitude is widely shared, in the self-help movement, in humanistic psychology, in transpersonal psychology, and in the relative new discipline of positive psychology. Having the proper mindset is also emphasized as a key to business success in numerous business workshops, seminars, and groups, including several I belong to, including BNI—Business Networking International; B2B—Business to Business Gathering, and BNF—Business Networking Formula. I have even written several books about having this mindset and have used this approach as a guide to my own life and business.

Commonly this positive outlook on life is described as leading to a more successful, satisfying, productive life, which includes having better relationships, more energy, and being less likely to fall ill. But the emphasis in these self-help programs and books is on current benefits, not on what one might gain in the future, even though some research studies have shown the relationship between one's attitude and longevity.

For example, Kekich describes how Harvard researchers followed 1306 men, who they rated as having high levels of optimism or pessimism, for 10 years starting in 1986. Over the course of the study, those men who were very optimistic had almost half the risk of suffering from any coronary problems compared to those who described themselves as very pessimistic. In another study by the Mayo clinics, the optimists had few physical health problems, felt more energy, felt generally more peaceful and happy, and had fewer problems at work or in their daily life.[34]

One's attitude also contributes to the placebo effect: because of their positive attitude that a drug will help them, patients who think they are receiving a drug often improve as much as those patients that actually receive it. Having the belief makes the difference because your thoughts and emotions influence your health. We don't only feel a certain way *because of* our physical condition; how we feel also *causes* our physical condition, a frequent theme in health, wellness, and self-help workshops and seminars.

People can stave off diseases and get better faster based on their beliefs. This power of positive thinking is a key element in the power of prayer—when people know others are praying for them, that helps them heal. Numerous books, such as Norman Vincent Peale's classic *The Power of Positive Thinking*, originally published in 1965, make this mind-body connection.[35] So again and again, there is support for the principle that your thoughts, emotions, and beliefs have a major influence on your physical health: you experience fewer illnesses and recover more quickly from any illness or injury, which affects how long you will live.

Having this positive attitude has enabled people to engage in amazing feats as well as to attract others—having good relationships is linked to a long life as is having a purpose and goals. For example, consider the case of Stephen Hawking, who was struck down with ALS, a neurodegenerative disease, when he was just 21. Even though he became wheelchair bound and increasingly unable to move his body, he has been able to survive by his will to contribute to science through studying the laws that govern the universe. Aside from his many awards and accomplishments, including being a Fellow of The Royal Society and a member of the U.S. National Academy of Sciences, Hawking has written numerous books— some by dictating his ideas or directing a computer with his thoughts, such as *A Brief History of Time* and *The Universe in a Nutshell*.[36]

Another example of this power of positive thinking is Nicholas James "Nick" Vujicic, an Australian Christian evangelist and motivational speaker, who was born without arms and legs. Despite some early struggles due to his disability, he learned to use his toes like fingers so that he could turn a page and operate an electric wheelchair, computer, and mobile phone. When he was 17, after his mother showed him a newspaper article about a man dealing with a severe disability, he began giving talks to his prayer group. A few years later, after he graduated from Griffith University at 21 with a B.A. in Commerce and a major in accountancy and financial planning, he founded a secular motivation speaking company based on the power of having the right attitude and wrote his first book, *Life Without Limits: Inspiration for a Ridiculously Good Life*, published by Random House in 2010 and translated into more than 30 languages. He also created a DVD for young people called *No Arms, No Legs, No Worries!* His positive attitude played a major role in helping him not only survive but also to become a great success as a motivational speaker.[37]

These examples also show the importance of having a strong sense of purpose. It gives you a sense of direction, meaning, and determination to do something. This purpose can be inspired by faith but it doesn't have to be. It can be related to a cause, to inspiring others, or to just having success in business. Whatever it is, having a purpose helps you feel good about yourself and what you are doing. There are many examples in the news of people who have lost their sense of purpose upon losing a job, losing a close partner or child, or experiencing a business failure. For some people, all is lost and they die soon afterward because their sense of purpose and positive, hopeful outlook is gone. But others find a way through the tragedy, often by using the experience to teach a lesson to others so that they have a new purpose to guide them forward.

What can you do to develop this positive, purposeful attitude that will contribute to your living longer? Here are some possibilities.[38]

- Use positive affirmations, which express something you want for yourself, whether a personal quality or an achievement, in the here and now. State it as if you already have it, and that will help make it happen.

- Remind yourself of your past achievements and successes, since this will help you feel more confident about what you can do to set what you want in the future.

- Expect that you can do what you want to do—and expect that the best things in life will come to you.

- Compliment and reward yourself for who you are and what you have done—and repeat these compliments and rewards on a regular basis, especially after you accomplish a goal.

- Envision your goals and prioritize them so that you go after what is most important to you and create realistic plans that will help you get what you want.

- Accept whatever comes and view anything bad that happens as a learning experience so that you can build on that and grow rather than feel regret and blame yourself and others for what has occurred.

- Forgive yourself or others for anything that has gone wrong so that you don't continue to bear a grudge or hold hate in your heart, which can undermine your health due to your negative thinking.

- If you fall ill or are injured, imagine yourself overcoming that illness or injury and visualize yourself sending healing energy to that part of your body. Imagine yourself overcoming that infection, believe you will get better, and know that you will. Also take your mind away from the pain to something pleasant, which will help you heal faster.

- See yourself aging more slowly or even becoming younger, and that vision will help slow or reverse the aging process.

In short, use your imagination to support your willpower. In this way you bring your unconscious, intuitive part of yourself to supplement your conscious, rational intention, which helps make what you have thought about become real. Use self-talk and visualization to give more power to whatever you want to do and that will help make that happen. By imagining yourself healthy and fit, you send a message to your body to act in ways that support that action, since thoughts have real-world effects. Those who embrace positive thinking typically have more success, satisfaction, and joy in everyday living. At the same time, whatever you do to get healthier and happier will contribute to your long-term health and survival.

Managing and Overcoming Stress

A closely related idea to the benefit of having a positive attitude is the emphasis on managing your stress and reducing it when you experience a stress overload. A first step to successful stress management is to recognize that there are two types of stress: a beneficial type called "eustress" and a negative type called "distress." The first type puts some pressure on you to work toward a goal or overcome a difficulty or challenge. According to James Lee of *The Methuselah Project*, we are "evolutionarily prepared" for this kind of stress: it involves experiencing a stress event followed by action and then a resolution. By contrast, distress involves being upset and even stuck by a problem, and usually when people talk about feeling stress, they are referring to distress.[39]

Thus, it makes sense to talk about "managing" stress rather than simply eliminating it from your life. You want to keep the good acute stress, which keeps you alert and motivated. It's a mechanism that motivates you to take on a challenge, solve a problem, or overcome a difficulty in your life. Such stress makes you stronger through the process of "hormesis," which in biology refers to experiencing a small dose of something that enables you to better resist something in the future. It works like a vaccine, where you are inoculated against a serious case of something by getting a small, nonlethal dose.[40] As the popular saying and song goes: "What doesn't kill you makes you stronger."

By contrast, avoid chronic, cumulative stress. While an individual stress or two might motivate you to do more or perform better at something, a continued series of stressors can prove to be too much, leading to sudden anger, ongoing depression, or turning to some outlet for escape. For instance, to relieve a buildup of stress, one might turn to alcohol, take drugs, or vent frustration and anger on others, such as by lashing out in road rage.

Too much stress is also associated with a number of biological changes and illnesses that shorten one's life. One change is the shortening of the telomeres at the ends of the chromosomes that occurs with aging and become even shorter with stress. For example, in *Life Extension Express*, David A. Kekich points out that the telomeres of people who feel more stressed are almost 50 percent shorter than those of people who feel less stress, which represents a 9–17-year difference in biological age. Kekich also notes that chronic stress can be a killer by "weakening your immune system; disrupting your digestive system; causing heart disease, stroke, cancer, Alzheimer's and more." Moreover, stress is a major reason people visit a doctor, since 80 percent of all visits to doctors in the United States are due to stress-induced conditions, and 90 percent of all diseases are the result of stress or made more serious by stress. Other conditions caused or acerbated by stress include high blood pressure, kidney damage, ulcers, food allergies, diabetes, and obesity. Also the main stress hormone, cortisol, causes the body's lean muscle to break down while you store more fat.[41]

This stress can be caused by a number of factors, including workplace pressures and life changes where you feel out of control. But one of the biggest causes of stress is when you react to what happens in your life by letting these events or others control you, rather than you taking control and influencing what happens in your life. Thus one way to manage is stress by taking control so that you plan what happens in order to move forward to your goals. Then, as much as possible, do what you can to delegate or outsource to others what you don't want to do. Another approach is to relax and calm down, using various techniques such as meditation, yoga, prayer, self-hypnosis, deep-breathing exercises, creative visualization, listening to calming music, biofeedback, or spacing out in a hot tub.[42]

Some of the other life extension activities already mentioned, such as exercise, can help in overcoming stress, while a good diet and supplements can help you better deal with the effects of stress. Such a diet and supplements will help reduce the increased production of free radicals due to stress, and exercise can increase the cardiovascular system's ability to handle stress by increasing the production of antioxidants. Many physical relaxation techniques, such as meditation and deep breathing, can reduce the effects of stress.

Meditation has life extension benefits as well. Many help decrease your biological age while helping you feel more relaxed. As Kekich notes, some of meditation's major benefits are:

- increasing the growth of new brain cells
- improving key mental skills, including comprehension, mental focus, memory, concentration, and decision making
- decreasing strength, anxiety, and depression
- reducing free radicals in the body and reducing blood pressure
- reducing the risk of stroke or heart disease
- speeding up weight loss[43]

In short, meditation as well as relaxation in general can lead to many physical, mental, and emotional changes that promote health and well-being; and numerous speakers, workshops, conferences, Meetup groups, and other organized programs offer different techniques for practicing these activities. The particular methods vary, such as by teaching different ways to breathe, focus on sounds or objects, or use images. But regardless of method, in practicing these techniques, the idea is to keep your attention focused on the activity of the exercise and ignore any external distractions. Additionally many other strategies to reduce stress are widely recommended by those writing self-help books or speaking to groups on personal development. Among the stress-reduction strategies recommended by many writers, speakers, and consultants are:.

- Spend some quiet time, whether you meditate, read, write, or otherwise seek to clear your mind.

- Spend some time with your family or friends to build your social connections.

- Let go of your feelings of frustration, anxiety, or fear by talking it out with others in your social network or a therapist.

- Control your feelings of anger, bitterness, and hatred; otherwise they can turn against you and undermine your health and well-being if you dwell on them, since your thoughts, emotions, and physical body interact. Take steps to release negative feelings.

- Decide on your priorities so that you can concentrate on what is most important first and you don't feel overwhelmed or scattered by all the things you have to do.

- Be flexible so that you can respond to pressures to change, both in each day and in your life.

- Eat better and exercise regularly, since these can help your body deal with the effects of stress.

- Remind yourself to maintain a positive outlook so that you can look on negative experiences as opportunities to learn and grow.

- Use self-examination to seek harmony and overcome inner conflicts that cause stress, then take steps to release any unexpressed emotions and resolve outstanding conflicts, since those are major sources of stress.

- Don't dwell on the bad things that might happen, since that might lead to them actually happening—such as when you fear you won't do well on a test or in a job interview—as well as creating unnecessary stress for yourself.

- Look for the positive in every negative situation to maintain balance in your life, since balance is "the natural order of the universe."

- Learn to accept the things you can't change so that you can move on and don't get stuck in the past. This change can involve new ways of thinking or doing things, which are particularly important in today's rapidly changing high-tech world.

- Don't let others' negative remarks bother you. Remind yourself that not everyone is going to like you or what you do; be ready to put the negative comments behind you and move on.

- Recognize that nothing is perfect and that it's better to get something done than to wait to do it perfectly. You can't achieve full perfection, so you'll feel a lot less stress if you don't keep trying and inevitably failing. Do the best you can and feel satisfied with what you achieve.

- Accept the praise for what you have done rather than think you didn't earn it or otherwise putting yourself down in the face of success.

Relaxation

Relaxation is another approach to dealing with stress, as described by Ray Kurzweil and Terry Grossman in *Transcend: Nine Steps to Living Well Forever*. As they explain, stress is a survival reaction to the "fight-or-flight" mechanism that derives from our human ancestry, since early humans had to deal with life-or-death situations by fighting or running away. Humans still respond to any threat of risk in this way, and this response triggers certain biological processes that prepare you to run or stand and fight.

Here's how the biological response works. The amygdala, the part of your brain that perceives danger, triggers the pituitary gland to release ACTH (the adrenocorticotropin hormone), which triggers the release of cortisol, the stress hormone from your adrenal glands. The result is that your muscles and brain have more energy, which makes your physical and mental abilities stronger. At the same time, your adrenal glands produce adrenaline and noradrenaline that suppress your digestive, immune, and reproductive systems, since you don't need these to address the immediate crisis. Meanwhile, your blood pressure, blood sugar, cholesterol levels, heart rate, and respiration increase so that you have more energy and ability to respond; fibrinogen levels increase so that your blood will clot more quickly in case of injury. Even your pupils dilate so you can "better see the threat or a means of escape."[44]

In other words, when you feel danger, whether caused by a physical threat or threat to your mental, emotional, or social well-being, you experience a series of biochemical changes that prepare you to respond to this threat. In ancient times, the physical exertion that followed these biological changes used up the excess energy. But today, when you repeatedly experience this flight-or-fight response to everyday stressors, you may respond by trying to not express your initial reactions, or you redirect your visceral reaction into a tamed, calmed-down response. But by doing this, you don't get the usual physical release, and that can lead to a series of serious ailments and illnesses. That's because you don't use up the hormones and fats that have been mobilized for action, while your overly high heart rate and blood pressure can trigger other negative effects (such as tension on the walls of the arteries, gastrointestinal disorders, cancer, heart disease, and stroke). Plus you can experience negative mental effects (such as feelings of anxiety, depression, problems with concentration, and insomnia) and compulsive activities (such as overeating, gambling, and drug use). The result is that experiencing chronic stress can accelerate the aging process.

What are these major sources of stress? Kurzweil and Grossman provide a list of the major stressors and the risk of getting sick from each one, using the Holmes-Rahe Social Readjustment rating scale. The incidents that are highest on the list include the death of a spouse, a divorce or marital separation, imprisonment, or the death of a close family member. It may be impossible to avoid all stressful events in daily life, and different things may be more or less stressful for different people. For example, a change in school, recreation, or religious or social activities may be relatively low-stress activities for most people but they could turn into very stressful activities for others.[45]

You can use a number of strategies to reduce harmful stress, as numerous writers, researchers, and speakers recommend. These strategies might include staying balanced in your work and personal life, knowing your priorities, having good interpersonal relationships, having a healthy diet, exercising, and taking a break from your daily routine.

But most of all Kurzweil and Grossman recommend the "relaxation response," which you can do each day to manage your stress. As they describe, this approach was first used in the mid-1970s by Dr. Benson and his associates at the Harvard Medical School and Beth Israel Hospital. They found that certain practices, such as meditation and yoga, are especially helpful in dealing with stress because they "produce physical and emotional effects that are opposite of the fight-or-flight response." Among other things, Benson and his associates found that these techniques can reduce a variety of biological functions triggered by this relaxation response, including lowering respiration, heart rate, blood pressure, and sugar levels. When used regularly, the relaxation response can also "permanently reduce blood pressure and enhance digestion, sleep patterns, mental acuity, blood-flow, and mood."[46]

To achieve these results, you can use a variety of relaxation techniques, including any number of forms of meditation, yoga, visualization, and biofeedback. The key is to practice whatever techniques you prefer on a regular basis. Avoid responding to stress with certain responses that can increase your risk of heart disease and your mortality rate from a heart attack, such as responding with an excessive level of anger, suspicion, and hostility. The danger to your health comes from responding to the ups and downs in life in negative ways so that your body's fight-or-fight mechanism remains in a nearly continual state of readiness. That readiness to respond negatively has negative effects on your body, makes your life more unpleasant and shorter, and increases the stress levels of others around you.[47]

Likewise avoid abusing drugs, smoking, alcohol, and caffeine as responses to stress, since they will also have negative effects on your body and mind. It's best to use different forms of relaxation when you are feeling stressed. You will not only feel better at the time since you will reduce the bodily effects due to stress, but also, as a result, you will add to your life span. In short, when you feel stressed, think of how you might relax, and you'll live longer as well as feel better.

PART III

The Potential for Immortality

None of these scientific, medical, or technological endeavors are practical yet, but still they offer a real potential for success, and several dozen billionaires and companies are pursuing these different approaches. Some of these efforts may combine for success in the future. For example, some are designed to preserve the brain, considered the seat of consciousness and the personality, so the brain could potentially be removed and equipped with a younger body. Other approaches are designed to rejuvenate the cells to stop or reverse aging, while other research is dedicated to eradicating certain diseases. In the following chapters, I briefly describe these strategies, none of which is ready for prime time. But where there is scientific commitment and research, there may be hope.

CHAPTER EIGHT

It Could All Be in the Brain

Related to faith-based efforts to claim that the soul or spirit (as distinguished from the mind or brain) is immortal—that the mind can exist separate from the brain—the spiritualist approach to scientific efforts has been conducted to show that the brain can survive separate from the body. This has been an argument the cryonicists have made in preserving the head through a neurosuspension, on the grounds that it can be reanimated on another body or exoskeleton when medical science has advanced far enough.

However, now there are other efforts to provide a much longer life or even immortality by separating the brain from the body. One strategy is to transfer the head to another body, presumably of someone who is brain dead. The other strategy is to transfer the brain—or a copy of the brain, with all its neurological connections intact like a piece of software—into a new human body, exoskeleton, or even a computer. These different strategies may sound like way out science fiction, but they are happening now. The fascination with this approach has even inspired several movies based on this science becoming a reality in the future. But within a few years, this effort could become very real.

I'll describe each of these approaches in turn.

Transplanting the Human Head onto a New Body

Plans to transplant a human head onto a new body have recently been in the works, gaining widespread media attention since August 2016. The plan is to do the highly complex operation in 2017, probably in the United Kingdom, assuming all preliminary tests go well and the scientists don't encounter legal and political hurdles that would prevent them from carrying out the operation.

The initial announcements about this revolutionary new procedure date back to April 2016, when *Newsweek* and other publications announced plans for this dramatic new medical operation occurring in 2017. As these publications proclaimed, Italian neurosurgeon Sergio Canavero was ready to perform a two-part human head transplant procedure in which he would remove a human head from one body and fuse it with the spinal cord of another body. He called the head transplant procedure "HEAVEN" for "head anastomosis venture," perhaps with a hope for acceptance by the spiritual and faith communities, and he called the spinal cord fusion "Gemini."[1]

Canavero actually announced his plans for the process three years earlier in a June 2013 paper in the peer-reviewed journal *Surgical Neurology International*. He then presented his plans in 2015 as the keynote address at the American Academy of Neurological and Orthopaedic Surgeon's 39th annual conference. But now the media was paying attention.[2]

His plan is certainly revolutionary, involving a long, complicated operation in a specially equipped hospital suite that will last 36 hours, cost $20 million, and involve at least 150 doctors, nurses, technicians, psychologists, and virtual reality engineers. The volunteer originally scheduled to participate in this procedure was Valery Spiridonov, a 31-year-old Russian program manager in the software development field (although now according to more recent articles[3] as of this writing in July 31, 2017 suggest the patient will now be a Chinese national volunteer to occur in December 2017). Spiridonov was selected since he suffers from Werdnig-Hoffmann Disease, a rare and often fatal genetic disorder that renders the body immobile, because the disease breaks down muscles and kills nerve cells in the brain and spinal cord that enable the body to move. As a result, Spiridonov's condition is much like that of Stephen Hawking, the noted British physicist, who suffered from the progressive generation of his muscles as a result of the ALS motor neuron disease. Like Hawking, Spiridonov is confined to a wheelchair, his limbs are shriveled, and his movements are limited to feeding himself, typing, and using a joystick to control his wheelchair.[4] He volunteered for the procedure in 2015 because he couldn't see any other way of continuing to live. He would rather take the chance of dying if the operation didn't work than continuing to live in his shriveled body. Canavero has predicted the operation has a 90 percent chance of success.

The operation itself sounds somewhat like the cryonics procedure that separates a head from its body by decapitation and then places it in a tank at a very low temperature. But the big difference is that the brain is expected to wake up after being placed on another body, where it is fused with the spine. Then the patient will be placed in a deep coma while the head-body connection heals.

More specifically, what happens in the operation is this. First, the patient's body will be placed in one hospital suite, while the brain-dead patient who is donating his body is placed in a nearby room. Presumably the donor patient will be a good match, since he has been matched with the Russian for height, build, and immunotype, increasing the chances that the patient will be more "at home" in the new body and reducing the chances that the body's immune system will reject the patient's head.

The operation will then proceed as follows. The surgeons will anesthetize both patients, fit them with breathing tubes, lock their heads with metal pins and clamps, and attach electrodes to their bodies to monitor their brain and heart activity. Next, the surgeons will cool the patient's head to about 12–15°C (about 53–58°F) so that he will become temporarily brain dead. Next, the doctors will drain the blood from his brain and flush it with a standard surgery solution, after which a vascular surgeon will loop tubes made from a silicone-plastic combination called Silastic around the carotid arteries and jugular veins. Then the surgeon will tighten these tubes to stop the blood flow and later loosen them to allow the blood to circulate again once the head and new body are connected. After this, both teams, working at the same time, will make deep incisions around each patient's neck and use color-coded markings to note all the muscles in the patient's head and the neck of the donor, so they can be connected properly.

After the bodies are prepared comes the main event—cutting through both spinal cords with an extremely thin diamond nanoblade while viewing this cut through an operating microscope. Operating in both rooms, the surgeons perform this cut together. Then, they rush to attach the patient's head using a custom-made crane with hook-and-loop straps to shift his head onto the donor body. They use clamps on the main blood vessels to prevent air from coming in to cause a blockage. To complete the attachment, the surgeons have to quickly sew the arteries and veins of the patient's head, after which they have to adjust the spinal cord stumps so they are even. Then they fuse together the spaghetti-like axons in the nerve cells of both the head and body using a polyethylene glycol (PEG) glue that enables the cytoplasm of the cells in the head and body to join together. This chemical has previously been found to promote the regrowth of cells in the spiral cord.[5]

The result, if all goes well, is a spinal fusion that links together the head and body. Then the surgeons will make assorted sutures and reconnections of various nerves and muscles, and they will start the blood flowing again, which they have to do within an hour in order to revivify the body with the patient's new head. At the same time, doctors have to use medication to suppress the body's immune system.[6] Needless to say, this is an extremely complex operation, in which over 100 doctors have to coordinate all kinds of procedures in the two bodies to make sure everything works.

Assuming the operation goes well, the patient will remain in a drug-induced coma for four weeks so that his brain can recover. While he is in that state, the doctors will regularly check his blood for antidonor antibodies, and they will electrically stimulate the spinal cord through implanted electrodes so that the neurons in the head and body will communicate with each other and the patient's sensory and motor functions will improve.

Finally, after about four weeks, the doctors will awaken the patient, assuming he is still alive, and he will start a rehabilitation program that involves training to adjust to his new body and to learn to walk again. A virtual reality training program that requires the use of bodily movements is already planned to begin a year before the patient undergoes the operation to prepare him for what he might experience when he wakes up with a new body.[7] Once he does, he will begin the process of learning to walk again. Presumably, if all continues to go well, as *Newsweek* reports, "Canavero predicts his patient will be able to walk three to six months after surgery."[8]

Yet, there could be many risks—from paralysis to death. For example, one potential for failure is whether the axons in the sections of the spinal cord that have been fused together will form a connection that works. Another possibility is that Spiridonov's brain could experience irreparable damage while it is cooled down and doesn't have any blood flow.

Still, despite the risks, Canavero is determined to go forward with the new Chinese volunteer. To that end, Canavero is hoping to perform the operation in the United Kingdom, if permitted to do so, although he has other countries in mind. He has already begun organizing the team, which so far includes a Chinese surgeon, Dr. Xiaoping Ren, to work with him.[9] And before they do the surgery on the patient, Canavero says he might use a fresh cadaver as a proxy to test out the many procedures.[10]

While this operation will be a first-ever procedure on humans, there have been some successes with animals, which Canavero has used to support the potential for the operation in humans. For example, scientists in China performed a head transplant on a monkey, and they were able to connect up the blood supply between the monkey's head and new body, although they did not attempt to reattach the spinal cord. In another case, researchers in South Korea and the United States claimed that they reconnected the spinal cords in mice and in a dog. In still another experiment, Dr. Kim and a team of researchers at the Konkuk University in Seoul severed the spinal cords of 16 mice and injected the polyethylene glycol (PEG) into half of the mice. Even though the mice who didn't receive the PEG died, five of the eight mice who did receive it regained some ability to move. Then, when the researchers gave an enhanced version of PEG to five rats with severed spinal cords, they found that electrical signals passed down the spinal cord after their treatment; they weren't able to test for the restoration of movement since four of the rats were killed in a flood at the laboratory. In the case of the experiments with the dog, 90 percent of its spinal cord was severed, and it was initially paralyzed. But after the researchers used the PEG solution, three days later the dog could move its limbs, and after three weeks it could walk and wag its tail.[11] So there is some basis to conclude from the results of the prior research on animals that the procedure proposed for the operation on Spiridonov might work on humans.

But will it? And should Spiridonov take the risk? Canavero's proposed operation has certainly met much skepticism from the scientific community, just as this reaction greets any revolutionary new scientific procedure before it is fully tested. For example, Dr. Arthur Kaplan, the head of medical ethics at NYU Langone Medical Center, claimed in an article in *Forbes* that Canavero's planned operation is "both rotten scientifically and lousy ethically." Dr. Jerry Silver, a neuroscientist at Case Western Reserve University, who has worked on repairing spinal cord injuries, called Canavero's proposed transplant "bad science," and he stated that "just to do the experiments is unethical."

A further argument against the procedure was raised by two Italian bioethicists, Anto Cartolovni and Antoni Spagnolo, who asserted that it was not possible to separate the brain from the body, as proposed by Canavero. In their view, according to modern cognitive science, human cognition is an "embodied cognition," in which the body plays an important part in the formation of the human self. Therefore, they thought that even if the operation were successful, the person with the new body would encounter "huge difficulties to incorporate the new body in its already existing body schema and body image," and that would have strong implications for defining the human identity.[12]

Yet Canavero would seem to have already taken that brain-body connection into consideration by providing for virtual reality training before the operation and by providing an extended retraining program after Spiridonov wakes up from the coma, in order to become "comfortable" in his new body.

In any event, apart from any issues a successful transplant might raise, such as the rights of any offspring—since they would have the genetic make-up of the donor, not the individual with the brain in the donor's body—the big question is, will it work? Or even if the operation doesn't work in Spiridonov's case, just as the first heart transplants didn't work, the operation might provide the groundwork for future successful operations.

Another consideration, as Canavero suggests, is that the operation might be a way to help individuals who are paralyzed gain mobility. Additionally, the operation has the potential to length the life of anyone whose body is failing by giving that person a new body from the neck down. That doesn't necessarily mean the person could live forever, since the brain can be subject to disease or neurological damage with age. There is always the potential to reverse any brain damage, since that is another type of research going on to increase longevity or the potential to live forever, as will be described in the next chapter.

As of late June 2017, plans are still going ahead for the first operation. The surgeons are waiting for permission to use a donated body from a young male brain-dead patient.[13] The operation is officially set to take place in December 2017.[14]

Transplanting the Brain into a New Body

The procedure for transplanting a brain into a new body is quite different from transplanting the whole head. It is a much more complicated process, and so far no scientific experiments on humans are planned. But there is a lot of theorizing about how this process might work, and a number of films since the 1960s have explored this possibility and the results—generally bad.

The reason this procedure is more complicated and far off in the future is that it involves transferring the nerve cells and all their connections in the brain or transferring the contents of consciousness, like transferring all the content and system files on a computer, but the brain is composed of far more "files." The key obstacle is being able to connect the nerve fibers from the transplanted brain to the spinal cord of the donor body.[15] On a theoretical level all of this may be possible, but it is in the distant future. Even so, some early developments may be laying the groundwork for a future brain transplant.

The basic theory is that a brain transplant, if it works, could be a way for a person who has an aging body or is experiencing organ failure to transfer his or her own personality, memories, and consciousness into a new, younger, functional body. Of course, this reintegration of a brain and a new body would only work if the person's brain was in top functioning condition. If the person was experiencing any neurological damage or disease, such as Alzheimer's or other dementia, he or she would not be a candidate for a brain transplant, although other medical technologies are being developed to overcome neurological disorders, as will be described.

Although no human brain transplant has yet been attempted, some research suggests the brain could be transferred. In one experiment, which was actually a head transplant, neurosurgeon Robert J. White grafted the head of one monkey onto another, whose head had been removed. Subsequent EEG readings showed that the brain in its new body was functioning normally, although the host's immune system attacked the brain after a few days, so the monkey died nine days later. If a head transplant should eventually prove possible as a way to transfer consciousness to another body, this research might contribute to the research effort on how to transplant the brain by itself.

The big barrier to making a brain transplant possible is that nerve tissue is not able to heal properly since scarred nerve tissue does not transmit signals well. But this problem could be resolved: some recent research at the Wistar Institute of the University of Pennsylvania has shown that tissue-regenerating mice, known as "MRL mice," can regenerate nerves without scarring. Also, in 1982, Dr. Dorothy T. Krieger, the chief of endocrinology at Mount Sinai Medical Center in New York City, successfully performed a partial brain transplant in mice. If researchers can make some breakthroughs in animal research, perhaps doctors might later apply this knowledge to making a human brain transplant.

The rationale for the operation is much like that of transplanting a human head onto another body—to enable someone who has a degenerative disease, like Stephen Hawking, or has total paralysis from an accident to have a healthy body.[16] Certainly, having a healthy body would enable someone having a brain transplant, if possible, to live longer. But the potential for staying alive forever in this way is limited, since the brain ages along with the rest of the body, even though it might do so at a slower speed for many people as the other body parts wear out. However, while it may be possible to replace other organs, such as the heart, lungs, and kidneys, with organs from human or animal donors or with artificial parts, it is not possible to simply replace the brain. As a result, for a person suffering from brain cancer, Alzheimer's, or other forms of dementia or neurodegeneration, it is only a matter of time before the disease progresses so far that the person can't be saved. Perhaps one day other medical breakthroughs may stop or reverse some of these diseases, as discussed in the chapter on biological developments to promote longevity. But until then, any brain transplant is limited by the health of the individual's brain.[17] As a result, even if a brain is successfully transferred into another body, the brain will continue aging and could be affected by any of these degenerative brain diseases, thereby limiting how much longer the individual can live in a new body.

Nevertheless, assuming the medical issues can be resolved so that the nerves can be successfully connected and any brain diseases conquered or delayed, what would the effect be on the person whose brain is transferred? Aside from adjusting to the new body, as in the case of the head transplant described above, the person could experience major personality changes, because, as Angelique Bordey, PhD, professor of neuroscience at Yale University School of Medicine, notes, "we grow our brain/mind to our body. So personality could change just due to the psychological shock."[18] Even so, the person's identity would be "more like the identity of the donor of the transplanted brain," according to Khalid M. Abbed, MD, professor of neurosurgery at Yale, since "the brain is where identity and personality are stored."[19] Thus having a brain transplant is not like getting a new heart or lung; the donor's identity and personality go along with the transplant, although the donor will experience some major personality changes due to having this new body. As summed up by Konstantin Slavin, MD, professor of neurosurgery at the University of Illinois in Chicago:

> I do believe the day will come when there will be a "whole body transplant"— so the brain of a person will be given a new body (and not the other way around). If this happens, the identity and personality of the original brain owner will be allowed to continue within [the] new physical body, natural or artificial. This will not keep the person alive indefinitely as the brains age and degenerate over time—and I would not be very optimistic about our ability to stop aging completely.[20]

In short, a brain transplant could be one way to increase longevity for someone with an aging body. It could also enable individuals suffering from degenerative diseases to have a healthy new body, which could contribute to their living longer. But a brain transplant is not a panacea for living forever, since brains can age and deteriorate too, although in the future, new medical technologies might be able to restore these aging brains to health. Once they do, the donor body might come from a human donor who has recently become brain dead but otherwise has a healthy body (such as someone who has died from drowning). Or the donor body could be an artificial body, not just one that provides replacement body parts and organs, as discussed in the next chapter.

Transplanting the Brain into a Computer Interface

The other direction in brain transplant theorizing is considering the possibility that a brain-computer interface can be developed, so the consciousness of the brain can be transferred into a computer or mechanical body, such as a robot. As will be discussed, mechanical replacement parts are increasingly being used to replace body parts, from arms and legs to hearts and lungs, so potentially a computerized brain could become a receptacle for human consciousness once the brain transfer technology develops to that point.

Some research even shows that it is possible to use commands from the brain without the need for the body, such as in one study by C. Ethier and associates reported in *Nature*. The researchers conducted a study with a monkey and showed that it was possible for the monkey to use commands from the brain and bypass the spinal cord in order to move his hand.[21]

In such a transfer of consciousness, the problem of brain aging would be overcome, since in this scenario the contents of the brain, including thoughts, memories, and beliefs—basically everything that makes up the human personality and identity—would be transferred, much like transferring data from one computer to another. Assuming this technology could work, presumably one copy of an individual would be transferred. But if a human being can be reduced to a very large computer file, what would prevent it from being placed into multiple computers that could assume multiple forms from desktops and laptops to robots? Would that mean it could be possible to create multiple identical humans?

So far this possibility of transferring the brain as computerized data is far in the future, since it involves carefully mapping the structures and connections within the brain, which has over 100 trillion connections. Furthermore, a specialized supercomputer would be needed to store this computerized brain and enable it to function. But despite that limitation, some scientists believe that these hurdles will be overcome, so downloading a brain will become possible and could be a way for a dying person to live on in this other form. And if this technology works, potentially anyone might want to use it when they have various health issues due to aging.

One such believer is Dr. Hannah Critchlow, a Cambridge University neuroscientist, who claims that if a computer could be built to recreate the

100 trillion connections in the brain, it would be possible for human consciousness to exist inside such a program.[22] Such a supercomputer is necessary because the human brain is composed of about 100 billion nerve cells. As such, it is "the most complicated circuit board" one can imagine, and it uses electricity to function. Whereas the brain's weight is only about 1.5 kilograms—about 2 percent of the body—the brain takes about 20 percent of the body's energy consumption in the form of this electricity. At any given time, the brain is only using a small percentage of its power, so it is effectively running in low gear. But when you engage in some mental activity, such as thinking about some topic, the brain revs up its power for that purpose.[23] Even so, any computer used to download the brain has to be powerful enough to download the entire brain, not just the portion of it that may be used for a time for a particular purpose.

Another well-known scientist who thinks a transfer of consciousness is possible is Stephen Hawking, who had this to say about the operations of the brain: "I think the brain is like a program in the mind, which is like a computer, so it's theoretically possible to copy the brain onto a computer and so provide a form of life after death."[24]

A key to this possibility is the development of a sufficiently powerful computer, which many scientists and thinkers now believe is possible. For instance, in 2005 the head of British Telecom's futurology unit postulated that rapid advances in computing power would make "cyber-immortality a reality within 50 years." To illustrate, he described how the PlayStation game system was evolving, since the new PlayStation was now 1 percent as powerful as the human brain, and it had achieved supercomputer status compared to where the system was 10 years ago. Given that trajectory of development, he opined that "PlayStation 5 will probably be as powerful as the human brain."[25]

Another believer in the brain transfer possibility is Ray Kurzweil, a computer scientist, inventor, futurist, and author of seven books, including *The Age of Spiritual Machines*. In 2013 he estimated that the ability to transfer the entire human mind to a computer would be achievable within four decades, based on increasing the amount of computation needed to simulate a human brain. In his view, this computer power would be able to increase a billionfold in that time, making the transfer of a human mind to a computer possible.

The underlying theory that we can recreate our brains on a computer is based on the notion that the human mind is essentially like a computer sending electronic signals in a program. Therefore, if we can decipher this code and develop a powerful enough computer to read this code and thereby match the 100 trillion connections in our brain, we can simulate consciousness. Then, we can download our consciousness into this computer program.[26]

Meanwhile, some initial developments toward this possibility of downloading the human brain have been done by researchers working with animals with much smaller brains. For instance, Henry Markam and his team conducted a study through the Blue Brain project in which they successfully simulated a rat's neocortical column, which is a complex layer of brain tissue common to all mammals. In conducting this simulation, they emulated the processing power of a brain with 1.6 billion neurons—about the size of a cat's brain.

However, the computer processing approach still has to overcome the problem that even with the best computer hardware that now exists, the computer doesn't have the actual processing power of a biological brain. Computers currently cannot process information in parallel by doing multiple calculations at the same time, whereas a biological brain can operate about 10–100 times as fast. For instance, whereas a cat might process information in a second, it might take the Blue Gene supercomputer from 10 to 100 seconds, depending on the complexity of the task.[27]

The other difficulty for creating the hardware needed to house a brain is the amount of storage space and processing power required. Given the vast number of neural connections in the brain—about 100 trillion according to the estimates previously noted—a complete map would require about 20,000 terabytes and require 1,016 FLOPS (floating point operations per second) of processing power to function. But researchers believe that level of computing power will be possible within a few years. As Jordan Inafuku and his colleagues at the Stanford Computer Center explain:

> Currently, only the world's fastest supercomputer possesses the capability of crunching that many numbers in a second. Nevertheless, the future is bright. If computing technology continues to follow recent trends such as Moore's law, doubling in processing power every two years, it is likely that most supercomputers will be able to run an accurate simulation of the human mind within the next few years. Additionally, as storage capacities continue to increase, it will be more feasible and economical to store digital maps of human brains.[28]

To illustrate the speed of this development, Inafuku and his associates note that in 2007, the largest cortical simulation only contained about eight million neurons, about half the size of a mouse brain. But four years later, scientists could emulate brains made up of over 1.5 billion neurons, about the size of a cat's brain.

Meanwhile, as the processing power of the hardware has been developing, other researchers are working with ways to get supercomputers to simulate the mind. For example, researchers working with artificial intelligence have been developing robots and other machines that can think, reason, and learn by mimicking the functions of the human brain. These researchers have even developed computer interfaces that can read the signals of the mind.[29] For example, using these interfaces, individuals have been able to trigger movements on a computer, such as selecting letters on a keyboard, just by thinking what they want the computer to do.

Another of the many neuroscientists who believes this brain to computer transfer is possible is Dr. Ken Hayworth, who similarly views the brain like a computer. In the view of Hayworth and many neuroscientists, the brain "turns inputs (from) sensory data into outputs . . . through computations," and these produce our behaviors. If neuroscientists can map the brain, it might be copied in a computer, along with the individual mind resulting from those inputs.[30]

Thus, unlike spiritualists who view the mind as able to exist free of the body, neuroscientists regard the mind as a creation of the various computerized operations in the brain that make you, you. As Hayworth explains, mapping the complex connections of all the neurons in the brain, called the "connectome," is the key to making this transfer possible, since the connectome encodes all of the information that creates who we are, even though Hayworth acknowledges this process isn't yet proven and is not yet possible with the existing technology. Plus there is another theoretical challenge to be overcome. Even if neuroscientists can create the exact wiring diagram of the human brain, they have to be careful that nothing goes wrong while uploading it into its new computer home. Accordingly, neuroscientists will have to carefully monitor the uploading process and watch how the neurons are performing.[31] After all, a technical glitch could make the brain operating system inoperable or perhaps turn you into someone (or something) else.

Still, the potential for mapping the brain's neural structure is the core principle driving the neuroscientists working on applying this approach, and like many of these neuroscientists, it doesn't matter to them whether this computerized "you" is housed in a physical body or in a computer simulation controlling a robotic body.[32] So maybe the sci-fi movies of the last few decades aren't so far off—such as imagining Hal, the humanlike talking computer, who first made his appearance in *2001: A Space Odyssey* (1968), or imagining a cop as a robot in *RoboCop* (1987), where a police officer, brutally murdered by a gang of criminals, reemerges as a superhuman cyborg law enforcement officer.

Nevertheless, despite these challenges, one 35-year-old Russian Internet millionaire, Dmitry Itskov, is investing in this hope for immortality, and he believes that within 30 years it will be possible for everyone to live forever. One reason for his fervent commitment to this technology is that otherwise he expects to be dead in 35 years, because for now death is inevitable. That condition exists, since as we get older, the cells making up our bodies lose the ability to repair themselves, so we become vulnerable to all the age-related conditions, such as cardiovascular disease, that kill about two-thirds of human beings. But if a brain upload becomes possible, that can bypass aging using "cutting-edge science to unlock the secrets of the human brain and then upload an individual's mind to a computer," which will free them "from the biological constraints of the body"—or as Itskov states: "The ultimate goal of my plan is to transfer someone's personality into a completely new body."[33]

In fact, some techniques by neuroscientists for mapping the brain are making this goal more and more possible. Some neuroscientists have recently been using magnetic resonance imaging (MRI), high angular resolution diffusion imaging (HARDI), and more invasive techniques to construct a detailed brain map. Besides mapping the anatomical structures, researchers have also used these methods to map the connections that link the various regions of the human mind/brain. Among other things, using functional and resting state MRIs, neuroscientists have been able to create accurate images of neuronal circuits that are activated under different simulated behaviors. They can also detect the fluctuations in the brain activity of people at rest, which gives them further information about the coordinated networks in the brain. And with the HARDI technology, they can measure how water diffuses along the fibrous tissue in the brain as well as visualize the long, slender nerve cell projections, called axonal bundles, that conduct electrical impulses within the brain. Plus, even more invasive techniques are uncovering details about how the synapses that link individual nerve cells to each other work. So far, researchers at Stanford's School of Medicine have already used these techniques to map mice brains by slicing portions of the cerebral cortex and staining each section to reveal a 3-D mosaic of synapses. They can then rotate and examine these synapses while preserving the original anatomical structure of each neuron. Neuroscientists think that these techniques can be applied to higher-level brain structures in order to map them, and eventually they can use this approach to create accurate models of the human brain.[34]

Once this modeling happens, neuroscientists believe it will eventually be possible to map the brains of specific individuals. Then, by using other technologies of the future, such as brain-computer interfaces, they can eventually develop the necessary link between minds and machines so that it becomes possible to upload the consciousness of a living human subject.[35]

However, this technology leads to many questions. If you can upload this complete consciousness into another body like a computer program, what happens to the original? Would one want to delete it because it's in the old deteriorating body? Is it possible to perform multiple uploads and then download those into other human or robotic bodies? What if something goes wrong with the upload? Would part of the upload and download process include creating a backup as in the case of a computer program or data? Or could the successful operation of one of the transplants mean that no other copy could operate at the same time, much the way that software companies incorporate certain blocks in the program, so it can only be downloaded once? What if something goes wrong with the programming in the original copy? Can this be replaced by a new uncorrupted copy, just as one might call a software company to request a replacement or new download? And what kind of medical professional or systems engineer would supervise the process of completing or replacing an upload? The possible ramifications of this method of seeking immortality through a re-creation of consciousness like a computer program raises all kinds of questions—not only about the physical operations of the process if this is possible but also of the ethical and legal issues that might arise, since the process taps into the fundamental question of human identity.

The Potential of a Brain Transplant Explored in Film

Given all of these issues raised by the prospect of a brain transplant, it is not surprising that this path for living longer and even forever has been embraced by sci-fi writers of both novels and films. They have created plots based on the assumption that a brain transplant is possible, but then something goes horribly wrong. While many of these sci-fi stories started off as novels later turned into films, many of the more recent tales were produced as films, often of the action-adventure variety. I even wrote a script for one sci-fi adventure called *Brain Swap*, with plans for it to be filmed in January or February 2018. I'll describe this literary and film genre briefly to illustrate how writers have considered the implications of what might happen if a brain transplant becomes a reality.

The beginnings of the genre go back to 1928, when Edgar Rice Burroughs, the author of *Tarzan of the Apes*, published *The Master Mind of*

Mars, sixth in the series about John Carter of Mars. In the novel, he imagines that Ras Thavas, the mastermind of Mars, has solved the problem of transplanting a brain, and he has applied this new technology to create a new form of business—selling young bodies to rich old people. To carry out this scheme, he has organized a network of slave buyers and kidnappers, who obtain young healthy bodies to sell to his clients. Then he takes the brain from a young slave and replaces it with the brain of his client, who thereby regains his youth.

Unfortunately, the mastermind of Mars is aging himself and does not trust any of his assistants, believing they might kill him in order to take over his position. So Thavas recruits U.S. Army Captain Ulysses Paxton, who has arrived on Mars and has no relation to any Martian, and instructs him in how to perform the operation. But plans go awry when Paxton falls in love with a beautiful slave girl whose body has been sold to a rich old woman, and he promises her to recover it, thereby undermining Thavas's evil scheme.[36]

Later, beginning in the 1960s, a series of mostly action/adventure and horror sci-fi movies were produced based on the brain transplant concept.

One of these was *Monstrosity*, a 1963 black-and-white Indie film, called *The Atomic Brain* in its TV release. Again the driving force is an aging wealthy person who wants a younger body. In this case an elderly woman uses her vast wealth to convince an eccentric and brilliant scientist to transplant her brain into a youthful body. She chooses one of three young immigrant women hired to be servants, and she makes arrangements to replace the young woman's brain with her own.

The transformation in *Change of Mind* (1969) also involves changing places with someone of a different race or social position. In this case, a white district attorney, David Rowe, has his brain transplanted into the skull of a black man to radically change his appearance. But once he does, his wife has difficulty adjusting to her husband's new appearance, and David now feels the pain of racial prejudice for the first time in his life. At first the local sheriff resents David, but after he kills his own black mistress, he has to turn to David for a legal defense. When David investigates the black woman's murder, his superiors, friends, and family treat him differently. Thus, *Change of Mind* is an early attempt to consider the potential for a brain transplant to upend traditional social relationships, when the transplant results in the brain ending up in a very different type of donor body. Usually the practice involves someone old seeking a new body, although other reasons might lead one to seek a body of a different race or ethnicity as an alternative to plastic surgery. For example, in a crime movie, a criminal seeking to become someone else to avoid prosecution after a crime or career in crime might want to totally transform his appearance.

In some films, the brain transfer just leads to carnage. For instance, *Brain of Blood* (1971) featured a brain transplant to save the life of Amir, the benevolent ruler of Khalid who is dying. Immediately after his death, he is flown to the United States, where a doctor transplants his brain into the body of a simpleton, but it is put in the wrong kind of body. The film then turns into a bloody horror movie, which includes a woman chained in the basement, as Amir's friend and his wife try to save him.

In *Who Is Julia?* (1986), the distinction between the wealth of the brain donor and the lower-class status of the person providing the body drives the plot: a beautiful wealthy woman is hit by a truck, and she is nearly killed. At the same time, a plain-looking lower-middle-class woman faints and suffers brain death. Since the brain of the beautiful woman is fine, the doctors transplant her brain into the "plain Jane's" body. But problems develop when the plain Jane's husband continues to believe she is still his wife, even though she has no memory of him. And why would she, since she now has the beautiful woman's brain? But when the plain Jane goes to resume the beautiful woman's life, she doesn't mix well with her new socialite's friends and family, since she now has the plain Jane's looks and suffers when she looks in the mirror and sees her new appearance.

Total Recall (1991), starring Arnold Schwarzenegger, based on the premise of transplanting memories for an interesting vacation, suggests that a brain transplant can become a popular pastime for the wealthy. However, as might be expected in a popular film, the transplant goes horribly wrong. As the film begins, it is the year 2084, and a construction worker, Douglas Quaid, is having horrible nightmares about being on the planet Mars with a woman who is not his wife. After seeing an ad on the subway for ReKall, Inc., a facility that implants fake memories of ideal vacations, Quaid decides to go on an adventure trip to Mars as a secret agent. But once he goes on the trip, he loses his real memory, discovers he is a former Mars Intelligence agent, and becomes involved with a woman working for the rebels. Soon assassins working for the colony's corrupt administrator are after him while he is fighting against the evil Mars administrator. Meanwhile, his real life wife turns on him, and he discovers his original identity has been erased and a new one implanted, which includes a woman who serves as his wife so she can watch over him. The plot becomes very convoluted, but in the end, he and the woman working for the rebels prevail and get rid of the evil administrator. But despite this success, he wonders if he is really having a dream and if all of these experiences have only been happening in his head back at Rekall. Or are these experiences real? Even though the film is mostly an action adventure, it raises the issue of identity when a brain—or the memories in one—is placed into a new body, raising questions about who one really is.

Sci-fi comedy *Man with the Screaming Brain* (2005) features a wealthy businessman whose brain is combined with that of a Russian taxi driver by a mad scientist. As the story begins, William Cole, the wealthy CEO of a U.S. drug company and a stereotypical ugly American, travels to Bulgaria with his wife Jackie in the hopes of diversifying his company's financial interest. After they are driven to the hotel by a taxi driver and former KGB agent, Yegor Stragov, Cole's wife secretly cheats on him with Yegor. Later, after Jackie sees Cole kissing the hotel maid, who tries to rob him, the maid hits Cole on the head with a pipe outside the hotel, and when Yegor sees this, the maid kills him too. After that, a vengeful Jackie has Cole's life support disconnected at the hospital, after which the maid kills her as well.

Then the brain operation occurs, in which a mad Russian scientist takes the damaged parts of Cole's brain and combines them with healthy tissue from Yegor's head. After Cole wakes up and escapes from the warehouse where the operation occurred, he can hear Yegor's voice in his head, and together they plan to get the woman who killed them. Meanwhile, Jackie's brain is put inside a robot, and after she escapes she too seeks revenge on the maid. Eventually, after more complications leading to the death of the maid, Cole begins to experience brain damage and dies due to the cells of both brains not being able to coexist in the same body, and Jackie's batteries finally die. But all is resolved in the end, since the mad scientist's assistant brings the bodies of Cole, Jackie, and the maid back. Eventually, Cole returns to the United States. while sharing his body with Yegor's brain, and he is back together with Jackie, whose brain is now in the maid's body.

More recently, in *Self/less* (2015), a billionaire business tycoon Damian Hale is dying of terminal cancer, when he finds a business card directing him to Professor Albright. Albright tells him about a radical medical procedure called shedding in which his consciousness can be transferred to an artificially grown healthy body. Hale opts to do so and learns how to walk, swim, and engage in other activities in this new body, even though he experiences vivid hallucinations, which Alright says are side effects of the procedure and gives him medications to stop them. Then Hale starts a new life in New Orleans with a new pseudonym, Edward Kittner. But after he forgets to take his medications, he has hallucinations about a woman and a child, and, convinced these are actual memory, he follows these visions to St. Louis, where he discovers that he has the body of the woman's deceased husband Mark. Eventually he learns that Mark sold himself to Albright to raise money to cure the couple's young daughter. Meanwhile, Albright's henchmen try to kill Hale, but he and Mark's wife and daughter escape, and then he discovers that the inventor of shedding has shedded himself into Albright's body. There are more efforts to kill him and more escapes, until Hale and Mark's wife and daughter end up in the Caribbean, where Hale stops taking his medicine and dies peacefully, whereupon the real Mark wakes up in his own body. Mark discovers a video message from Hale thanking him for the time he gave him, and Mark reunites with his family at last. So again, there is the conflict between who rightfully has the identity after a brain transfer, along with all kinds of conflicts arising out of the former life of the donor who died and the person whose brain was transferred into the donor's body.

The notion of a human brain transferred into a cyborg's body is the basis of the *RoboCop* series, which features a 2014 remake of an earlier 1987 feature. In the remake, set in 2028 Detroit, Alex Murphy, a loving husband, father, and good cop experiences a critical injury while on duty, and a multinational corporation decides to create a part-man, part-robot police officer. But as the RoboCop seeks to deal with crime like a superhuman law cyborg law enforcer, he is haunted by recurring memories from his past.

More recently, in *Criminal* (2016), an American sci-fi action-crime thriller, a convict is implanted with the memories of dead CIA agent Bill Pope in order to finish his assignment. This assignment is to find out where the CIA agent hid the Dutchman, a hacker who has created a wormhole program that allows the owner to bypass all the computer codes protecting the world's nuclear defense codes. But the CIA agent has gotten killed after getting the Dutchman to a safe house.

That's where the brain transplant comes in. The agent's supervisor contacts a doctor, Dr. Micah Franks, who has developed a treatment that can theoretically plant the memory patterns of a dead man into a living one. So Franks keeps the agent's brain stimulated to preserve its knowledge and selects a convict, Jericho Stewart, to be the recipient, because Stewart had a childhood brain trauma that left him with underdeveloped front lobes, limiting his emotional development and impulse control. But here too, complications ensue, when Jericho escapes custody after the operation. After faking his death, he heads to the CIA agent's house, where he meets Pope's wife. As the action revs up and both Pope's killers and the CIA try to catch up with Jericho, he gradually has more and more flashes of Pope's past experiences. Eventually, after he is able to save Pope's wife and her daughter, they end up on the beach where Pope and his wife had their honeymoon. After Jericho meets Pope's family and they confirm that some part of Pope exists in him, he finds happiness with Pope's wife, and Pope's supervisor might even offer him a job. In this case the donor of the brain ultimately prevails, although (as often happens in these films) by adapting to or reconciling with the life that the owner of the body left behind.

Still another entry in this genre is *Ghost in the Shell* (2017), based on a 2015 Japanese series. The basic story is that in the near future, Major Motoko is "the first of her kind: a human saved from a terrible terrorist attack, who is cyber-enhanced to be a perfect soldier devoted to stopping the world's most dangerous criminals," as described in IMDB's plot description.[37] What this means is that Major Motoko's brain is transferred into a cybernetic body having greater strength and ability. In such a cybernetic body, an automatic control system is created through a nervous system and brain, a mechanical-electrical communication system, or a combination of the two. In the original Japanese version, the setting is mid-21st century, where many members of the public possess cyberbrains, which combine their biological brain with various networks. The level of cyberization varies from simple interfaces to almost completely replacing the brain with cybernetic parts. These can be combined with various prosthetic body parts, so a person can obtain a complete prosthetic body to become a cyborg, which is what happened to Major Motoko Kusangi, due to a terrible accident as a child, so she has had a full-body prosthesis since then. However, cyberization has a set of risks, since such a brain is open to attacks from very skilled hackers, and the most dangerous ones can bend a person to their will, much like a hacker can obtain control over someone's computer today. The American version has taken over this basic plot structure, as the American Major Motoko Kusangi, played by Scarlett Johansson, participates in an antiterrorism squad and fights to save the world from the most dangerous criminals while avoiding dangers from the hackers who seek to control her.

Finally, I want to mention my own film in this genre to be filmed in August or September 2017, called *Brain Swap*. It tells the story of a once-wealthy businessman, Sam, who is losing everything—his business to a bankruptcy, his wife to a divorce, and his health to terminal cancer. He decides, with the help of his associates (a lawyer, CPA, and doctor), to have his brain transferred into the body of a younger man. But instead of the recently deceased donor expected, there is a mixup at the hospital, and he ends up in the body of a bank robber who has just died. As a result, when Sam gets this new body, the police think he has escaped, and he becomes a wanted man. Two years later, Sam, now Jack, has a successful new life in a new town, where he is a real estate agent with a new wife. Then he rescues a boy from drowning in a lake, and it becomes a big news story, whereupon the wife of the criminal arrives and extorts him for money to keep his secret. But eventually the story comes out, leading to a big trial over who he really is—Sam, Jack, or someone new. At the end of the trial, Jack is found not guilty because his body has a new brain; so now all of the creditors of Sam appear and sue him. The story concludes with Jack and his wife enjoying life in the Mexican paradise where they have escaped, since he now considers himself Jack, no longer Sam. Once again, the story features a wealthy man obtaining a younger body, but in this case the film presents a more realistic portrayal of what it might be like to become someone new and seek a new reconstructed life.

Thus, unlike the other procedures for extending life or seeking immortality, this brain transplant procedure, if it works, raises all kinds of issues about personal identity. Who is this new person? Supposedly the brain donor gains a new lease on life, but often the brain donor is affected by the original persona of the donor of the body. Sometimes that influence occurs because their brains have been combined, as in the *Man with the Screaming Brain*; because they have only had a partial memory implant, as in *Total Recall*; because the person has forgotten to take the necessary medicine to prevent memories of the body's donor from surfacing, as in *Self/less*; because the person finds it difficult to go back to a previous lifestyle due to being in a different body, as in *Who Is Julia?*; or for other reasons. Thus, there is always a difficulty in melding the two components of the self, even after a brain transplant operation is successful, in those works of fiction that have considered the implications of such an operation.

In turn, these fictional quandaries foreshadow the kinds of issues that could occur if these operations are successful in the future. In fact, one reason for this difficulty of completely separating the person with the donor brain from the person with the donor body is that consciousness is embedded within the body, not just the brain, as reflected in some of the concerns raised by scientists commenting on the head transplant operation. In their view, the transplant of a head onto a new body will not completely turn the brain donor into a person with a now-healthy body. Instead, assuming a successful fusion of the head and spine in the new body, the person with the brain will continue to show some qualities of the body donor because of the embedding process. How much? That is for future research to discover.

In the meantime, the speculations of fiction writers, including myself, raise these issues about identity that become salient when the brain of a person who has recently died is transferred into the body of a person who has become brain dead. Perhaps the potential for uploading a brain into a computer or robotic body might avoid this issue, since there will be no transplant, just copying the brain's neurological and cellular components to create a new functioning brain in another form. But then that brings its own set of issues, such as whether this is a true copy, whether there could be other copies, and what happens if a successfully uploaded brain later malfunctions.

These are all issues to be considered down the road as medical and computer technologies progress to where a brain transfer or transplant is actually possible. And if a transfer or transplant doesn't appeal to you, there are still other options for living forever through technological and biological developments to be described in the final chapter. None are ready for prime time, but there is extensive funding for research in these areas, so there is always hope.

CHAPTER NINE

The Biotech System for Longevity or Immortality

Finally, there are a number of biotech efforts to develop new solutions to enable humans to live much longer, if not forever. Many of these strategies have been funded by Silicon Valley billionaires, who are hoping for their own immortality by setting up research foundations to pursue different strategies.

The two main approaches are these:

- *the medical-technology solution*, involving new ways to create organs or body parts that can be surgically implanted in or connected to the body
- the *biotechnology strategy*, based on finding ways to reverse aging through the use of stem cells, cell rejuvenation, genetic manipulation, and antiaging medications

The medical-technology solution is an extension of current surgical procedures for transplanting hearts, kidneys, lungs, and prostheses for arms, legs, hands, and other body parts. By contrast, the biotechnology approach involves developing mostly new breakthrough methods. Many of the researchers have used experiments on lower forms of animals from jellyfish and hydras to mice to try different antiaging methods to enable the animal to live longer. Then they consider how the techniques that work on animals might be applied to humans.

I'll briefly describe each of the major approaches in turn, since these are all in the experimental stage, and a detailed discussion of the medical and technological procedures used in each approach would become very technical for a general audience. Additionally, a detailed analysis of the state of art in each area could be a book by itself, aimed primarily at a science and academic audience.

An Overview of the Major Antiaging Strategies

In the last few years, the interest in finding biological and technological solutions has grown exponentially due to a number of factors. Besides the interest of a longer-living older population, a number of researchers have found some promising early results, suggesting that a much longer life and even immortality are possible due to new findings by medical and biotech researchers. Moreover, about a dozen billionaires, primarily in the Silicon Valley, have contributed millions of dollars to fund different lines of research. The reports of these different lines of research have been promising, and those seeking to live longer can pursue multiple strategies. This way if one method doesn't work, such as taking antiaging medicines or having an infusion of younger blood, maybe another will, such as having stem cell injections or undergoing some genetic manipulation.

The Support from the Wealthy Billionaires

A number of very wealthy tech executives are contributing to the search for immortality, since they want to buy their way out of the inevitability of death, believing their money can help find a high-tech solution to do just that. They are approaching the quest to live forever with the confidence in founding their high-tech empires. So they have funded their institutes and other longevity sciences, with the view that "the traditional pace of science moves too slowly." They think that major government and educational institutions, such as the National Institutes of Health, require too much consensus and are not willing to take the needed risks that result in major breakthroughs.[1]

These investor billionaires might be compared to the entrepreneur-philanthropists at the turn of the 20th century, as Ariana Eunjung Cha points out. At that time, a small group of wealthy industrialists, led by Andrew Carnegie and John D. Rockefeller, made major financial contributions to create major change in the world by setting up schools, art museums, and public libraries in keeping with their ideals of democracy and equality. Today, tech titans are exercising their influence at a time of worldwide growing inequality, a time when the richest 1 percent of the world's population controls over 50 percent of the world's wealth, according to a 2016 Oxfam report from the World Economic Forum.[2]

The number of billionaires participating in this current effort to improve health and increase longevity reads like a roll call of the richest people on the planet. Some on this list and their areas of giving are:

- Microsoft: Bill and Melinda Gates: infectious diseases, child and maternal health
- Oracle: Larry Ellison: aging
- Amazon: Jeff and Mackenzie Bezos: immunotherapy for cancer research and neurological disorders
- Facebook: Mark Zuckerberg and Priscilla Chan: local health services
- Google: Sergey Brin and Anne Wojcicki: Parkinson's disease and genetics
- Microsoft: Paul Allen: brain research, artificial intelligence, and cell biology
- eBay: Pierre and Pam Omidyar: social technologies to improve health and well-being
- PayPal: Peter Thiel: aging and regenerative medicine
- Napster, Facebook: Sean Parker: allergies, cancer, and malaria
- AOL: Steve and Jean Case: traumatic brain injury and brain cancer
- Netscape: Mark Andreesen and Laura Arrilaga-Andreeesen: emergency medicine

An underlying belief of these tech titans is that the computerized analysis of large amounts of data can help in finding cures, predicting outbreaks, and discovering patterns that contribute to disease and aging that the human brain can't discover on its own. So they eschew the traditional scientific approach of beginning with a hypothesis, testing it with experiments, spending years refining and analyzing the results, and seeking corroborations from other scientists through peer reviews before publishing the results. Instead, they look for results from mining and mapping the huge sets of digital information obtained when people engage in various online activities. And today almost every online activity is tracked. Among these activities are searching for data, texting others, interacting on social networks, shopping, visiting doctors, and leaving a digital footprint wherever you go during the day. Then, with these billions and trillions of data bytes gathered, a supercomputer can test trillions of possible hypotheses at once to look for patterns and correlations that suggest solutions for major medical problems as well as analyze the role of thousands of genes in affecting everyone's health.[3]

In turn, a major focus of their contributions is not only on improving health and conquering disease for more healthy living now but also for attaining a longer life, or even immortality, through the advances in medical technology. To this end, these wealthy titans are funding four major pathways to immortality:

Avatars and artificial brains, discussed in the previous chapter. Among the scientists being funded is Russian scientist Drmitry Itskov, who believes immortality can be achieved by the year 2045; his website, 2045.com, seeks money from the wealthy in return for the end of death.[4] A visit to his website indicates that he has almost 44,000 members, and his website features the latest high-tech developments that show promise in promoting longevity, such as a March 8, 2017, article on how researchers have taken further steps to creating mind-controlled robots, and a March 10, 2017, report of a new study that shows that brain activity continues after death.

Ending death by disease. This funding is primarily earmarked for biotech companies seeking to cure cancer, all viral diseases, and other diseases associated with aging. Among the wealthy supporters funding these projects are Brian Singerman, a venture capitalist and a partner in the Founders Fund, and his partners, who include Peter Thiel and Sean Parker. These three men are not just interested in funding the development of drugs that extend life for a few months or years or that make living with a disease easier. They are seeking absolute cures, and they believe this is very possible. For instance, Singerman believes that "within 10 years all viral disease will be curable, and within that same time frame we will have a clearer understanding of what aging is, what causes it, and how to begin to stop it."[5]

Genetic tinkering. Here the wealthy are funding efforts to genetically turn off the aging process by employing some of the methods used by scientists to extend the life of lower animals, such as a 2001 study with roundworms.[6] In this research, scientists added an extra gene, known as SIR2, to the cells of roundworms, which lived for three weeks instead of the usual two. Perhaps this one-week increase may seem like a small amount, but if this approach were to work with humans, that would result in increasing the usual life span by a third (e.g., by 20 years if the average person lives to 60).

Cryonics. Already discussed at length, this is another search for immortality that the wealthy, such as Peter Thiel, are supporting. He has contributed several hundred thousand dollars each year to Alcor, based on the cryonicists' belief that medical technology will advance much further than in the past several centuries, healing both cellular and molecular damage and restoring full physical and mental health.

Some of the major wealthy donors and their organizations to reverse or stop aging are the following.

Peter Thiel. One of the biggest names in the Silicon Valley, Peter Thiel, has donated $3.5 million to the Methuselah Foundation, whose most visible spokesman is Aubrey de Grey, one of the cofounders. With an annual budget of $5 million (about $600,000 coming from Thiel), De Grey's efforts are devoted to the Strategies for Engineered Negligible Senescence (SENS) Research Foundation, which de Gray established in 2009. The foundation is devoted to finding drugs that cure the seven main types of age-related damage. According to de Grey, these are "loss of cells, excessive cell division, inadequate cell death, garbage inside the cell, garbage outside the cell, mutations in the mitochondria, and crosslinking of extracellular materials."[7] The underlying concept is that the human being is a machine with a structure that affects all of its functions; if scientists can find a way to restore that structure at the molecular and cellular level, functions would be restored, and, thereby, rejuvenation of the body would be achieved.

Project Calico. An even bigger immortality research behemoth, led by Sergey Brin. This project, founded in 2013, represents Google's effort to find a cure for death, and to this end, the company is going to contribute billions to a partnership with the pharmaceutical company AbbVie. This partnership was set up to find a way to reverse engineer the biology that controls lifestyle and develop methods to enable people to live longer and healthier lives, in part by developing drugs to stop or reverse the aging process. One goal is to create a drug to mimic the fox03 gene, which is associated with an exceptionally long life span.[8] So far Brin has contributed about $50 million to curing diseases such as Parkinson's, after he discovered through a genetic test that he was at risk of developing the disease.[9] As of the end of 2015, Google's parent company Alphabet had invested about $730 million into California Life Company (Calico), which has been partnering with universities and companies to develop pharmaceuticals for creating commercial products that extend one's life.[10]

Glenn Foundation for Medical Research, started by venture capitalist Paul F. Glenn in 1965. Since 2007, the foundation has annually distributed the Glenn Awards, which have provided $60,000 grants to independent researchers working on antiaging projects. In addition, the foundation has sought to initiate antiaging projects in large institutions, including Harvard, MIT, the Salk Institute, and the Mayo Clinic. It has annually contributed grants of more than $1 million through the American Federation for Aging Research, which is a charitable foundation that seeks cures for age-related diseases.

The Glenn Foundation also works with the Ellison Medical Foundation, founded in 1997 by Larry Ellison, the founder of Oracle, to contribute hundreds of thousands of dollars in grants each year to researchers seeking remedies for aging, totaling about $45 million a year. While many biomedical researchers focus on seeking cures for individual diseases, the

Ellison Medical Foundation is interested in looking for the root of the aging process in order to prevent aging and thereby stave off the many diseases that occur due to aging, most notably chronic diseases such as heart disease, cancer, stroke, and Alzheimer's. These diseases have become more widespread as people have come to live longer, currently with an average life expectancy of about 80 years, compared to about 47 years a century ago.[11] Most notably, the foundation has provided support for research on mice, which might someday be applicable to reversing aging in humans. So far Ellison has contributed about $6 million to the SENS Research Foundation, which is seeking to develop antiaging drugs.[12]

Another Silicon Valley supporter is hedge fund manager Joon Yung, who launched a $1 million prize challenge in 2014, in which he challenged scientists to "hack the code of life" in order to increase the human life span beyond 120 years—considered the maximum length of life (since the longest confirmed life span was 122 years). With input from a group of nearly 50 advisers, including scientists from some of the top universities in the United States, Yung offered his first prize for restoring the vitality and extending the life span in mice by 50 percent, and as of January 11, 2015, 15 scientific teams entered the competition.[13]

Other Companies and Organizations Created to Pursue Antiaging Strategies

In addition to the individuals already mentioned who are contributing money to support antiaging efforts, a number of other companies and organizations have emerged to pursue research or create antiaging drugs and services for the growing market of individuals seeking a longer life or immortality.

Besides Calico, Google set up Google Ventures with $2 billion in assets, headed by Bill Maris, the company's president and managing partner, and gives the company $300 million in new capital each year. While the company has stakes in over 280 start-ups, a big push is investing in technology and medicine, such as a research project by Google X to create a pill to insert nanoparticles into the bloodstream to detect disease and cancer mutations.[14] Another project, Foundation Medicine, is using genetic data to create products to diagnose cancer. One such product is its Interactive Cancer Explorer, which is like a Google search for genetic information about patients that helps devise better treatments for them. Another Google-backed company is DNANexus, which is building a global bank of genomic information that uses cloud computing, enabling people around the world to use this bank in doing research.[15]

Another company, Human Longevity Inc., was established in March 2014 by Craig Venter, biotechnologist, biochemist, geneticist, and businessman (one of the first to sequence the human genome) and by Peter Diamandis, a tech entrepreneur who founded the X Prize Foundation, which offers $5–$30 million in prizes for solving modern-day problems. Their goal in setting up this company was not to compete with Calico or to create antiaging drugs but to focus on the genetics of aging by creating a giant database of 1 million human genome sequences by 2020, including from supercentenarians, in order to mine this data to learn what factors lead to a longer, healthier life. The idea is to be the central research center for those working in the longevity field.[16] Also, the institute plans to offer commercial genome-sequencing services to companies such as 23andMe and other companies that analyze human genetics. It has even set up partnership arrangements with health insurers to provide a discount to individuals who have their genome sequenced.[17]

The Development of Antiaging Medicine

This interest in longevity has also led to the emergence of antiaging medicine as a medical specialty, devoted to the "early detection, prevention, treatment, and reversal of age-related dysfunction, disorders, and diseases."[18] This practice is based on a health care model that "promotes innovative science and research to prolong the healthy life span in humans."

This specialty had its origins in 1992, when a group of physicians and scientists came together to discuss the implications of newly developing discoveries in identifying the mechanisms that led the body to deteriorate and become more vulnerable to age-related diseases. The group was formalized as the American Academy of Anti-Aging Medicine (A4M), based on the premise that "diseases and disabilities of human aging are largely preventable, treatable, and perhaps even reversible." As of 2005, thousands of physicians in both private practice and teaching hospitals around the world were practicing this approach to medicine. The total membership, which includes scientists, academicians, and government and university-affiliated officials, was 14,500 in more than 78 nations. Now, with further advances in the field, there are 26,000 members in 120 nations, including members of the general public.

While researchers are typically focused on one line of antiaging research, the field has embraced four major biotechnologies for extending the healthy human life span, which include:

- regenerative medical innovations that regrow damaged or diseased tissues and organs
- stem cell technologies to develop a supply of human cells, tissues, and organs that medical professionals can use in providing acute emergency care and in treating chronic, debilitating diseases

- genetic engineering advances, whereby scientists can change one's genetic makeup to eliminate diseases
- nanotechnologies, which enable scientists to use very small tools to manipulate and change the biology of a human[19]

From 2001 to 2005, scientists in the United States and other countries made notable progress in reversing aging using these approaches. A few standouts were these:

- Immature neural stem cells were introduced into the brain and matured into adult brain cell by researchers at the McKnight Brain Institute. This was a major breakthrough in that it was the first procedure to successfully replicate the process of brain maturation with much more precision than ever before. The process involved freezing the stem cells until needed and then beginning a cell-generating process to produce thousands of new neurons— a procedure with great potential for curing or stopping the deterioration of the brain in many diseases, such as Parkinson's and Huntington's disease, by regenerating new healthy parts of the brain.[20]
- Researchers from the University of South Florida used stem cells from human umbilical cord blood in rats, and they found it reduced the rats' heart attack damage. They injected the stem cells into the rats' hearts after they induced heart attacks. They found that the stem cells greatly reduced the amount of damage, and afterward the rats' hearts began pumping again almost normally.
- Dr. Woo Suk Hwang and researchers at the Seoul National University used the same technology they had used to clone the first human embryo for research purposes—and in one research project they created embryonic stem cells from nine patients with spinal cord injuries, diabetes, and a rare immune disorder. As a result of their work, they showed that embryonic stem cells could be derived from these patients through a nuclear transfer, and that they could use stem cells to cure various diseases, since stem cells could now be matched to patients based on their medical conditions.[21]

These reports of scientific achievements illustrate the kinds of results achieved by medical researchers over ten years ago, in the early days of antiaging medicine as a discipline. No wonder the billionaires and other investors saw the huge potential of these biomedical technologies to increase longevity in humans.

Following are some of the major biotech strategies. Even though they have been briefly noted before, here I want to describe these strategies in a little more depth.

Fixing the Cells and Molecules

One effort is finding drugs to cure the seven major types of age-related damage to the cells, the approach being explored by the Methuselah Foundation, cofounded by Aubrey de Gray, with funding from Peter Thiel, the Silicon Venture Capitalist, who has given over $3.5 million to the foundation. The seven types of damage are these:

- loss of cells
- excessive cell division
- inadequate cell death
- garbage inside the cell
- garbage outside the cell
- mutations in the mitochondria
- crosslinking of the extracellular matrix

The underlying premise of this approach is that the human body is like a machine with a structure that determines all aspects of how it functions. Therefore, if it is possible to restore that structure at the molecular and cellular level, it will be possible to restore its function. Therefore, the body can be rejuvenated with the aid of the necessary drugs that repair the cells and molecules in the body.[22] The process is like fixing a car by repairing the parts so that they function like new.

Reversing Aging with Younger Blood

Another approach, which may sound a little vampirish, is using the blood of a younger donor to replace the blood of an older recipient to reverse aging.[23] The beginnings of this idea to transfuse blood to make older people more youthful dates to as early as 1615, when a German doctor and alchemist, Andreas Libavius, proposed connecting the arteries of an old man to those of a young man. He imagined that the young man's blood would energize the blood of the old man like a fountain of youth, but it is not certain the procedure actually happened. In 1660, when the Royal Society was first founded in London, some experiments in blood transfusion occurred to prolong life by replacing old blood with new.[24] Unfortunately, these early transfusion experiments proved deadly, since the scientists didn't know about blood groups or coagulation. As a result, the procedure was soon banned, at first in France and then in England.

Even the pope weighed in by endorsing the bans in 1679. Experiments with blood transfusions did not occur for the next 400 years.

Then, in 1956, Clive M. McCay, a gerontologist at Cornell University, revived the idea by experimenting on live mice. He sewed the flanks of live rats together: one was lively, healthy, and young; the other was old and in failing health. He described his attempts in the *Bulletin of the New York Academy of Medicine*: the results seemed promising in that the old rats—sometimes called "mice" in reports about his work—seemed to "age in reverse, getting healthier and younger as the experiment continued," while the young rats seemed to age prematurely.[25] The experiment didn't always work as intended, since if the two rats weren't adjusted to being connected, one would chew the head off the other. However, the positive results in many cases inspired scientists to conduct similar experiments, and they got similarly positive results. These early efforts were abandoned in the 1970s.

In 2004, Amy Wagers, of Harvard University's Department of Stem Cell and Regenerative Biology, revived McCay's flank-stitching experiments. She found the procedure worked, so with some of her research funded by the Glenn Foundation and Ellison, she sought to determine the individual proteins in the mouse blood responsible for the young-blood effect. She discovered a protein called GDF11, which was common in the blood of the young mice but was present only in small amounts in the bloodstream of the older mice, suggesting that its lack of presence was related to aging in the older mice. As she further found, GDF11 helped keep the stem cells, which are responsible for tissue renewal, active, so that healing occurred more quickly. By contrast, with a drop in GDF11 levels, healing was slower and other signs of aging developed.[26]

Meanwhile, some researchers looked for other proteins. Tony Wyss-Coray, a professor of neurology at Stanford University, who had been doing research on Alzheimer's disease using mice that were genetically modified to develop the disease, became interested in whether the molecular structure of human blood might reflect the state of the brain as it aged and deteriorated as a result of the disease. He brought together an international team of two dozen scientists to determine whether human blood did show this aging effect in the brain by analyzing the blood plasma from over 200 patients with Alzheimer's disease and comparing them to the plasma in healthy people. Eventually his effort to develop a test to diagnose Alzheimer's, even before individuals developed it, led him to observe that in healthy people the levels of certain proteins in the blood, which contains about 700 different proteins, fell with age. Conversely, other proteins increased and even doubled or tripled in old age.[27]

Several years later, Wyss-Coray teamed up with one of his students, Saul Villeda, to test out the idea of taking the blood from young mice, removing the blood cells, and injecting the plasma into the old mice. The experimenters would then compare the differences between the two mice by running a water-maze test to see if the old mice became more youthful, while the younger ones experienced earlier aging. To conduct the test, Villeda injected the plasma in each mouse every 3 days for 24 days. He used plasma from three-month-old mice, equivalent to humans in the 20s, and injected this plasma into 18-month-old mice, who were like humans in their 60s. The results provided dramatic proof for Wyss-Coray's theory, since the old mice with the young plasma proved to be aces at the water-maze test, and like mice half their age, they quickly raced back to the cage where they previously got an electric shock.[28]

Eventually, after additional testing, it appeared that the young plasma had an effect on a protein called CREB, which is a master regulator. This wasn't the only protein affected, since Amy Wager's study had shown that GDF11 was another rejuvenating protein in young blood. Villeda later found, as reported in a July 2015 paper,[29] that a second protein factor, B2M, is higher in the blood of old mice as in old humans, but when injected into young mice, their memories are suddenly impaired.

The results of these different lines of research all suggest the same conclusion—that certain substances in the blood contribute to aging. As noted by Ian Sample, "Among the hundreds of substances found in blood are proteins that keep tissues youthful, and proteins that make them more aged." According to Wyss-Coray's hypotheses, the reason for this result is:

> When we are born, our blood is awash with proteins that help our tissues grow and heal. In adulthood, the levels of these proteins plummet. The tissues that secrete them might produce less because they get old and wear out, or the levels might be suppressed by another genetic programme. Either way, as these pro-youthful proteins vanish from the blood, tissues around the body start to deteriorate. The body responds by releasing pro-inflammatory proteins which build up in the blood, causing chronic inflammation that damages cells and accelerates aging.[30]

So what does all this mean for human aging? To find out, in January 2014, Wyss-Coray and an associate, Karoly Nikolich, an entrepreneur and neuroscientist, established Alkahest, a company designed to separate plasma into its different parts and combine them into a "potent, rejuvenating cocktail." The plan was to first take human plasma, divide it into fractions rich in different proteins, and test them in mice to see if they increased brain function. Then, those that did so would be introduced into human trials, and those that work will be developed into a series of products to be made available to Alzheimer's patients.

A series of double-blind trials were conducted by Sharon Sha at the Stanford University's School of Medicine with at least 18 patients who were 50–90 years old and had mild to moderate Alzheimer's symptoms. In the trial, each patient received a unit of young human plasma or a saline solution once a week for four weeks. Then, after six weeks with no treatment, they had four more weeks of infusions; whether they got the plasma or the saline was switched from what they originally received. During the trial, doctors looked for cognitive improvements, and after the trial ended in October, 2015, Sha began analyzing the findings.

If the treatment works, there are some 15 million Alzheimer's patients around the world that might benefit from it; however, the entire world's plasma supply is only enough for half a million patients. So who gets it? That could be a question of ethics, on how an antiaging treatment is administered, although the most wealthy will probably be the ones who benefit first.

In any case, the first question is to see what works. The questions about the ethics of treatment can be considered later.

Fixing the Telomeres

Another biomedical approach has been to seek ways to restore the ends of the telomeres, which are caps at the ends of the chromosomes (like handles at the ends of a rope). As researchers have found, the telomeres become shorter with age. As Isaacson describes in an article on major projects that the Silicon Valley billionaires are backing, when individuals are young, an enzyme called telomerase keeps the telomeres healthy and stable, but as individuals age, their levels of telomerase drop, so the telomeres shorten, and the chromosomes begin to fray.[31] As a result, many researchers believe these fraying chromosomes might cause some of the physical declines due to aging.

The telomeres might play this role, since genes that make up all living organisms are twisted around the double-stranded DNA molecules called chromosomes, and the telomeres are at the ends of these chromosomes. These telomeres contribute to the division of these chromosomes, but each time a chromosome divides, the telomeres get shorter and shorter. But once they are too short, the cell cannot continue to divide, and after a while, these chromosomes die so that the cell can no longer replicate. Unfortunately, once cells are old or damaged, they can't be replaced.[32]

This replication process is critical, since the spiral DNA molecule has to split in half and reassemble a copy of itself, a little like a snake might get rid of its old skin. The telomeres at the end of the DNA molecule help protect the molecule from being copied incorrectly, resulting in some of the DNA code getting lost.

This copying process might be compared to copying a written document, in which the printed text might be compared to the DNA and the white space around it to the telomeres. Each time a copy is made, errors of alignment might be introduced, resulting in less and less white space and some lost text. This is much like what happens when replicating cells don't replace themselves properly, and the diminishing telomeres result in more and more errors each time.[33] As *Huffington Post* blogger Natalie Kalin notes, these differences associated with aging are dramatic, since a newborn has about 8,000 pairs of telomeres, whereas the elderly have only around 1,500 pairs.[34]

In other words, the fewer the telomeres, the more cell damage can occur, and as older cells deteriorate, they are more subject to attack by oxides and free radicals in the body and environment, which also contribute to aging. An early study by scientists at the Geron Corporation and at the University of Texas Southwestern Medical Center in Dallas seemed to support this. They tested the hypotheses that telomere shortening is the molecular clock that triggers the senescence that occurs when normal human cells undergo a certain number of cell divisions and ultimately stop dividing. In their tests, they compared two normal human cell types with and without telomerase, and they found a causal relationship between telomere shortening and cellular senescence, suggesting that stopping this shortening could contribute to life extension.[35]

As they explained in their conclusion:

> Our results indicate that telomere loss in the absence of telomerase is the intrinsic timing mechanism that controls the number of cell divisions prior to senescence.
>
> Cellular senescence is believed to contribute to multiple conditions in the elderly that could in principle be remedied by cell life-span extension in situ.[36]

Their breakthrough study led to further research on how to maintain the telomerase enzyme to prevent the shortening of the telomeres as well as looking for other factors besides telomerase that could contribute to this shortening process.

For example, Dr. Ronald DePinho, a senior scholar in aging at the University of Texas MD Anderson Cancer Center, and his colleagues investigated how these telomeres might play a role. They changed the genes of mice to create a strain of mice whose output of telomerase could be turned on and off, and they found that the enzyme played a major role in aging. When they turned the enzyme off, the mice got old much more quickly than usual, so in human years, they were 90 years old. The results were dramatic. The mice that aged prematurely had "shrunken brains, impaired cognition, infertility, thin bones, hair loss"—in effect, all the signs of an old mouse. But when the scientists turned the telomerase back on, the mice began to become young again. As DePinho reported, "The brain increased in size, cognition was improved, fertility was restored, hair returned to a healthy sheen, and all of the other problems that we saw in the animal were allevi-ated."[37] In effect, the aging process was not just stopped; the mice became younger, much like GDF11 had done in the research by Amy Wagers.

However, aging cells that don't have telomerase are more likely to become cancerous. Once the cells become cancerous, they have an increase in telomerase, so they spread rapidly. Thus, adding telomerase to an already aged individual could contribute to a growth of cancer, even though DePinho and other researchers believe that telomerase therapy might reduce the development of cancer, if the telomeres on the chromo-somes are less likely to shorten.[38]

Will this approach work in humans? The early results on mice are promising, although it is likely to be several years before these tests are done on humans. At the same time, if telomeres are one of the keys to aging, this opens up other research possibilities to learn what affects the increase or decrease of telomerase and what other factors besides telomer-ase might contribute to shortening or lengthening the telomeres.

For example, in looking for keys to suppress cancer, scientists at the Berkeley Lab, led by Martha Stampfer and James Garbe, looked at the cel-lular senescence process by which cells stop dividing. To this end, the scientists introduced molecular agents into the cells that contribute to cancer development in human mammary epithelial cells. One of the agents they introduced was a molecule called c-Myc, which reactivates the telomerase enzyme that maintains telomeres. This research led to a procedure that produces immortal human epithelial cells, without requir-ing any genomic alterations to the cells. As a result, they were able to study the mechanisms underlying immortalization in cells with normal genomes, so they could better understand the progression of cancer in order to develop new ways to intervene therapeutically in the earliest stages of cancer progression.[39] Because cancer is one of the age-related diseases, stopping its progression is a way to increase longevity.

Other Cellular and Molecular Factors Contributing to Aging

But is telomere length the only (or most important) factor contributing to aging? Kalin and other writers and researchers have pointed to still other factors on the cellular and molecular level. Two factors are oxidative stress caused by oxidants and glycation caused by glucose binding with DNA, proteins, or lipids. As Kalin describes it, the oxidants are reactive substances containing oxygen that damage the DNA, proteins, and fats. The older we get, the more we are exposed to these oxidants; thus our body suffers more damage.[40] One strategy has been to develop a series of drugs to reduce the oxidants that cause this damage.

In the case of glycation, which has a similar effect as oxidative stress, the DNA, proteins, and lipids are damaged when glucose, the primary sugar for energy by our body, binds with and damages these components. The problem becomes even worse as we age, so we experience more malfunctions throughout our body. Again, research on lower animals has helped show that oxidative stress and glycation are major culprits in aging. For example, in one study where researchers exposed worms to two substances that neutralized oxidants, the worms' life spans increased by about 44 percent.[41]

In another study, reported in *Nature* in 2011, scientists dyed worms yellow with a pigment called Thioflavin T (or Basic Yellow 1) and found that the worms lived 60–70 percent longer than usual. In another study, researchers found that the use of antifungal microbes called "rapamycin" increased the life expectancy of mice by 30 percent or more. Researchers have also found that spermidine, a molecular compound found in both grapefruit and human semen, has increased the life span of worms, fruit flies, and yeast.[42]

Still other studies have shown that certain drugs reduce the chances of dying from certain diseases. For instance, some researchers have found that rapamycin extends the life of mice and protects against cancer; other researchers have found that resveratrol, a compound found in red wine, reduces the risk of experiencing heart disease.[43]

Such discoveries have led to the development of a number of products that provide the missing stem cells, molecules, enzymes, and other ingredients associated with reducing aging. For example, there are now all kinds of longevity creams, capsules, and pills to increase longevity, such as Clustered Water (developed to rejuvenate structures within or between the cells) and Rejuvity's Ageless Renewal Serum (designed to revitalize the skin).[44] My own search on Google revealed dozens of nostrums, including Stem Cell 100 from Life Code, which revitalizes adult stem cells and reportedly doubles the normal life span of an animal model. As the website announces, "Rejuvenate your body and slow the aging process to help you feel and function more like a young person. This can help you feel better, look younger and improve your health." Another website, created by Longevity Science, features dozens of products selling at about \$25–\$35 each for a bottle of pills with medical names such as "Acetyl-L-Carnine" and "Ubiquinol."

Other products are designed to provide antioxidants, the nutrients in certain fruits and vegetables, that appear to protect the body from free radicals that cause cellular damage. But while some experiments have shown antioxidants have this curative power, other experiments have questioned whether oxidative damage is the main cause of damage. Even so, numerous manufacturers offer antioxidant-based vitamin supplements and other health products, claiming they have an antiaging effect, although it isn't clear if these supplements really do reduce oxidative damage to the body or really have an effect on aging.[45] Even so, you can find antioxidant superfoods in most supermarkets and health stores.

Do these products work because research on lower animals suggests they will work on humans? Do they work because of the placebo effect, so just believing in their powers will make someone feel better? Or do they work because they are able to heal key molecular components, resulting in positive changes in the cells or molecules in the body? It's hard to tell, because at this point most of the research on the biology and chemistry of aging is still preliminary. Even though testing has produced multiple positive results in lower animals, it is premature to conclude if and to what extent these findings apply to humans. Maybe they will; maybe they won't, but for now, we don't know anything for sure.

It is also difficult to distinguish the placebo effect in these products from the effects of the chemicals in them since one's beliefs interact with the endorphins. As Gollner points out, "stem-cell activating and resveratrol-laden age-management nostrums may not help prevent visible signs of aging, but it doesn't really matter if they work. What matters is that consumers *believe* they will work."[46]

Thus, while much of the research with lower organisms has been subjected to scientific peer-reviewed scrutiny, much of the research that is product driven (in order to come up with new antiaging drugs) is not. In his article "Live Forever! Can Science Deliver Immortality," Gollner raises some skeptical concerns that are still with us today. For instance, he points out that "nothing much supports the claims made by most eye-cream manufacturers," even billions of dollars are annually spent on cosmetic wrinkle-reduction treatments. Gollner also points out the dangers associated with human growth hormone (HGH), which is used as an antiaging remedy or for age-related problems, but these are not authorized by the FDA. Even so, HGH is used covertly by professional athletes, bodybuilders, and others, who falsely believe it can make people younger.

However, the misuse of HGH actually has the opposite effect, since it appears to cause organ malfunctions and tumor formation in test subjects. As a result, it increases the probability of an early death, the exact opposite of its intended effect in extending life. Still, HGH can be bought online and is promoted in a number of antiaging books, such as *Grow Young with HGH: The Amazing Medically Proven Plan to Reverse Aging*, by Dr. Ronald Klatz, who describes himself as "a world recognized authority on preventive medicine and advanced biotechnologies."[47] Klatz is also the president of the American Academy of Anti-Aging Medicine (A4M), previously cited.

For the moment, though, the scientific knowledge about aging and how to prevent it through methods to attack the cellular and molecular aging process is uncertain. The controversy currently comes down to the opposing views of the experimental genetic biologists and the evolutionary biologists. On the one hand, the genetic biologists, many with ties to pharmaceutical companies, believe we can chemically and genetically manipulate the longevity of genes and molecules and turn these findings about what manipulations are effective into medicines for humans. On the other hand, the evolutionary biologists don't think we can effectively impact the human aging process.[48]

Thus, while many scientists have found financial rewards through aging research, the findings are inconclusive, and the effects of these drugs on aging is still elusive. Nevertheless, the research goes on, and perhaps one day there may be more definitive results on how scientists and doctors can intervene to slow and eventually end the aging process—or maybe not.

New Body Parts and Organs

If the other methods of rejuvenating your cells or genes don't work, there is still hope in creating new parts of your body and organs. This approach may not be as dramatic as transferring your head or your brain onto a new human body or exoskeleton, but you can add in new parts to replace those that don't work, so you can live longer—and perhaps long enough for a more radical upgrade. Consider this approach a little like fixing your old car with new parts, from new wheels and hubcaps to getting a new paint job and doors to finally trading the old one in for a brand-new car. Here you are just fixing the car so it runs well again.

Already there are many operations that work and are widely known. You can get a new heart, liver, kidney, or lungs; a prosthesis to replace your arms and legs; plastic surgery to help you look younger and other surgeries to help you reduce your weight, such as a tummy tuck. In effect, you can already add a collection of new body parts and organs to create a whole new you.

Another method can help you replace a failed organ (such as a kidney or liver) when one isn't readily available via organ donation. As Isaacson describes, with the help of the Silicon Valley billionaires who are funding much of this research, you can now get a 3-D-printed liver or kidney. Doctors are also able to turn skin cells into stem cells and grow them into organs. A new procedure called a cold saline resuscitation is now available to repair many types of body damage. It involves replacing a patient's blood with an infusion of cold saline to drop the body's temperature, so the person goes into a state of suspended animation. In this state, doctors can repair many damages to the body that would otherwise be fatal, such as if the patient has been shot or stabbed, experienced a hemorrhage, or had an organ fail. It helps doctors make the repair if they already have the needed replacement parts on hand—either parts that have been cloned from the patient or produced in two ways: they can be grown in the laboratory, such as from stem cells, or they can be printed.[49]

To use the car analogy again, what if the number of replacement parts needed soars as a person gets older? At some point, one may want to upgrade to a newer and better model, such as a robotic or holographic avatar, made possible by advances in robotics, neural interfaces, and artificial organ creation, which is the goal of Itskov's 2045 initiative. This option is becoming more and more possible, since teleoperated robotic avatars already exist. These are robots that are controlled remotely by a human being through signals sent through a wire or a local wireless system (such as a Wi-Fi network, over the Internet, or by satellite). In Itskov's view, these avatars will eventually become superior in their abilities compared to the human's physical body, so avatars will become increasingly popular.[50] When this day comes, the individual can effectively live in an avatar outside the limits of his or her biological body, even though one's consciousness will still reside in one's brain. So one might imagine a scenario where humans who are incapacitated, like Stephen Hawkins, might have their own avatars that can move around and interact with others as if the individual were no longer wheelchair bound.

Later, even more developments in externalizing one's consciousness into another body may be possible, such as by creating a superpowerful exascale computer that can operate at the speed of a human brain, a development that Intel is already working on, with a goal of having it operational by 2018. Some seeds of this development were planted in August 2013, when Japanese and German researchers were able to simulate 1 percent of all human brain activity by using Japan's K supercomputer. Now, given that breakthrough, the next step is scaling up to represent even more of the whole brain at the level of the individual nerve cell and its synapses—all becoming possible by the new generation of exascale computers now under development.[51]

Notes

Introduction

1. Betsy Isaacson, "Silicon Valley Is Trying to Make Humans Immortal— And Finding Some Success," Newsweek.com, March 5, 2015. http://www .newsweek.com/2015/03/13/silicon-valley-trying-make-humans-immortal-and -finding-some-success-311402.html.

2. Isaacson.

3. "Juan Ponce de León," History Channel, n.d. http://www.history.com /topics/exploration/juan-ponce-de-leon.

4. Laurie-Anne Vazquez, "How Science Is Making Immortality a Reality," *Fiat Physica* (blog), April 6, 2015. https://www.fiatphysica.com/blog/making-history /science-of-immortality-aging-wrinkles.

5. Isaacson.

6. Douglas Perry, "Want to Live Forever? New Research Suggests Science Could Extend the Typical Human Lifespan Indefinitely," OregonLive, January 12, 2015. http://www.oregonlive.com/living/index.ssf/2015/01/want_to_live_forever _new_resea.html.

7. John Martin Fischer and Benjamin Michell-Yellin, *Near Death Experiences: Understanding Visions of the Afterlife*, Oxford: Oxford University Press, June 1, 2016.

Chapter 1

1. Ray Kurzweil and Terry Grossman, "Bridges to Life," in Gregory M. Fahy, Michael D. West, Michael D. West, L. Stephen Coles, and Steven B. Harris, eds., *The Future of Aging*, Norco, CA: Springer Science & Business Media B.V., 2010, p. 22.

2. Kurzweil and Grossman, p. 4.

3. Mark A. Lane, Julie Mattison, Donald K. Igram, and George S. Roth, "Caloric Restriction and Aging in Primates: Relevance to Humans and Possible

CR Mimetics," *Microscopy Research and Technique* 59 (November 6, 2006), pp. 335–338.

4. Kurzweil and Grossman, p. 4.

5. Kurzweil and Grossman, p. 5.

6. Kurzweil and Grossman, pp. 6–7.

7. Kurzweil and Grossman, p. 7.

8. S. N. Blair, H. W. Kohl 3rd, R. S. Paffenbarger Jr, D. G. Clark, K. H. Cooper, and L. W. Gibbons, "Physical Fitness and All-Cause Mortality: A Prospective Study of Healthy Men and Women," *Journal of the American Medical Association* 262 (1989), pp. 2395–2401.

9. Kurzweil and Grossman, p. 8.

10. Kurzweil and Grossman, p. 11.

11. Joel D. Wallach and Ma Lan, *Immortality*, 2nd ed., Bonita, CA: Wellness Publications, 2011, pp. vii–viii.

12. Wallach and Lan, pp. ix–x.

13. Wallach and Lan, pp. x–xi.

14. Wallach and Lan, pp. 3, 17.

15. Wallach and Lan, pp. 20–21.

16. Wallach and Lan, pp. 24–25.

17. Wallach and Lan, p. 32.

Chapter 2

1. "Top 10 Ways You Could Live Forever," Listverse, January 1, 2013. http://listverse.com/2013/01/01/top-10-ways-you-could-live-forever.

2. Madhumita Murgia, "Will Technology Help Us Live Forever?," *The Telegraph*, January 21, 2016. http://www.telegraph.co.uk/technology/2016/01/25/will-technology-help-us-live-forever/.

3. Larry Schwartz, "4 Ways the One Percent Is Trying to Buy Their Immortality," Alternet, June 12, 2015. http://www.alternet.org/personal-health/4-ways-one-percent-trying-buy-their-immortality.

4. Murgia.

5. Joel Lee, "Want to Live Forever? 6 Technologies That Could Stop Aging," MakeUseOf, August 12, 2015. http://www.makeuseof.com/tag/want-live-forever-6-technologies-eliminate-aging/.

6. "Top 10 Ways You Could Live Forever."

7. A. G. Bodnar, M. Ouellette, M. Frolkis, S. E. Holt, C. P. Chiu, G. B. Morin, C. B. Harley, J. W. Shay, S. Lichtsteiner, and W. E. Wright, "Extension of Life-Span by Introduction of Telomerase into Normal Human Cells," *Science*, New Series 279 (January 16, 1998), pp. 349–352.

8. Betsy Isaacson, "Silicon Valley Is Trying to Make Humans Immortal and Finding Some Success," *Newsweek.com*, March 9, 2015.

9. "Top 10 Ways You Could Live Forever."

10. Alexandra Grunberg, "From Cyborgs to Nanobots: 5 Ways Scientists Hope to Achieve Immortality for Humanity," Outerplaces.com, July 20, 2015.

https://www.outerplaces.com/science/item/9395-from-cyborgs-to-nanobots-5
-ways-scientists-hope-to-achieve-immortality.

11. Grunberg.

12. Murgia.

13. Lee.

14. American Friends of Tel Aviv University, "Turning Off 'Aging Genes,'"
ScienceDaily, January 2, 2014. https://www.sciencedaily.com/releases/2014/01
/140102123403.htm.

15. Stanislaw Burzynski, "Practical Application of Gene Silencing Theory of
Aging: Life Extension in Animal Testing and Human Clinical Trials," *Anti-Aging
Medical Therapeutics* XI (January 2009), pp. 447–454.

16. Murgia.

17. Lee.

18. Grunberg.

19. Murgia.

20. "Top 10 Ways You Could Live Forever."

21. Grunberg.

22. Lee.

23. Isaacson.

24. Mat Smith, "Japanese Latest Humanoid Robot Makes Its Own Moves,"
Engadget, July 30, 2016. https://www.engadget.com/2016/07/30/japan-humanoid
-alter-robot.

25. Lee.

26. "Top 10 Ways You Could Live Forever."

27. Schwartz.

28. Isaacson.

29. Isaacson.

Chapter 3

1. Bob Nelson, *Freezing People Is (Not) Easy: My Adventures in Cryonics*,
Guilford, CT: Lyons Press, 2014.

2. Robert C. W. Ettinger, *The Prospect of Immortality*, Painesville, OH: Ria
University Press, 2005, p. 11.

3. Ettinger, pp. 20–22.

4. Ettinger, pp. 20–31.

5. Ettinger, pp. 32–39.

6. Ettinger, pp. 36–41.

7. Ettinger, pp. 42–43.

8. Ettinger, pp. 50–51.

9. Ettinger, pp. 53–62.

10. Ettinger, pp. 66–67.

11. Ettinger, pp. 69–70.

12. Ettinger, pp. 79–84, 91–93.

13. Ettinger, pp. 94, 99–100.

14. Ettinger, pp. 102–103.
15. Ettinger, pp. 108–114.
16. Ettinger, pp. 119–123.
17. Ettinger, pp. 124–135.
18. Ettinger, pp. 140–143.
19. Ettinger, pp. 149–157.
20. Nelson, pp. 22–24.
21. Nelson, pp. 30–39.
22. Nelson, pp. 42–48.
23. Nelson, pp. 53–64.
24. Nelson, pp. 66–69.
25. Nelson, pp. 70–73.
26. Nelson, pp. 79–80.
27. Nelson, p. 82.
28. Nelson, p. 88.
29. Nelson, pp. 91–100.
30. Nelson, pp. 113–120.
31. Nelson, pp. 121–123.
32. Nelson, pp. 126–134.
33. Nelson, pp. 137–138.
34. Nelson, pp. 141–143.
35. Nelson, p. 143.
36. Nelson, pp. 146, 150–151.
37. Nelson, pp. 157–163.
38. Nelson, pp. 174–187.
39. Nelson, p. 187.
40. Nelson, pp. 188–189.
41. Nelson, pp. 188–192.
42. Nelson, p. 199.

Chapter 4

1. Tim Urban, "Why Cryonics Makes Sense," WaitButWhy, March 24, 2016. http://www.waitbutwhy.com/2016/03/cryonics.html.

2. George Dvorsky, "Generation Cryo: Fighting Death in the Frozen Unknown," Gizmodo.com, September 22, 2016. http://gizmodo.com/generation -cryo-fighting-death-in-the-frozen-unknown-1786446378.

3. "Generation Cyro: Fighting Death in the Frozen Unknown."

4. "Generation Cyro: Fighting Death in the Frozen Unknown."

5. Michael Perry, "Our Finest Hours: Notes on the Dora Kent Crisis," Cryonics, September, October, November 1992. http://www.alcor.org/Library/html /DoraKentCase.html.

6. Perry.

7. Ashley Collman, "Revealed: Red Sox Legend Ted Williams Was Cryogenically Frozen AGAINST HIS WILL by His Son in the Hope He Could Be Brought Back to Life Someday," *Daily Mail*, December 1, 2013. http://www.daily mail.co.uk/news/article-2516442/Red-Sox-legend-Ted-Williams-cryogenically -frozen-AGAINST-HIS-WILL.html.

8. David McCormack, "We Did It Out of Love: Baseball Legend Ted Williams' Daughter Finally Speaks Out about Why She and Her Brother Spent $100,000 to Have Their Father's Body Cryogenically Frozen," *Daily Mail*, May 19, 2014. http://www.dailymail.co.uk/news/article-2632809/We-did-love-Baseball -legend-Ted-Williams-daughter-finally-speaks-brother-spent-100-000-fathers -body-cryogenically-frozen.html.

9. Richard Sandomir, "Fight over Williams' Frozen Body May End Soon," *New York Times*, September 26, 2002. http://www.nytimes.com/2002/09/26/sports /fight-over-williams-s-frozen-body-may-end-soon.html.

10. Larry Johnson, *Frozen: My Journey into the World of Cryonics, Deception, and Death*, New York: Vanguard Press, 2009.

11. Johnson, p. 9.

12. Johnson, p. 21.

13. Johnson, pp. 20–22, 28.

14. Johnson, pp. 56–57.

15. Johnson, p. 86.

16. Johnson, pp. 93–97, 100.

17. Johnson, p. 125.

18. Johnson, p. 129.

19. Johnson, pp. 134–136.

20. Johnson, pp. 139, 145–146.

21. Johnson, pp. 154–158.

22. Johnson, pp. 159–166.

23. Johnson, p. 170.

24. Johnson, pp. 186–187.

25. Johnson, pp. 188–192.

26. Johnson, pp. 196–198, 204.

27. Johnson, p. 205.

28. Johnson, pp. 214–219.

29. Johnson, p. 221.

30. Johnson, pp. 223–224.

31. Johnson, pp. 231–233, 243–250.

32. Johnson, p. 259.

33. Johnson, pp. 265–269.

34. Johnson, pp. 271–275.

35. Johnson, pp. 281–283.

36. Johnson, p. 308.

37. Johnson, p. 350.

38. Johnson, pp. 350–351.

39. "Response to Larry Johnson Allegations," Alcor, n.d. http://www.alcor.org/press/response.html.

40. "Response to Larry Johnson Allegations."

41. "February 10, 2012: Lawsuits against Larry Johnson End; Johnson Issues Statement," Alcor, n.d. http://www.alcor.org.

42. "Problems Associated with Cryonics (and Some Possible Solutions)," Alcor, n.d. http://www.alcor.org/problems.html.

43. "Problems Associated with Cryonics."

Chapter 5

1. David A. Kekich, *Life Extension Express*, Newport Beach, CA: Booksurge Publishing, 2009.

2. Kekich, p. 7.

3. Kekich, pp. 13–19.

4. Kekich, pp. 23–24.

5. Kekich, pp. 25–28.

6. Kekich, pp. 31–32.

7. Kekich, p. 35.

8. Kekich, pp. 31–38.

9. Kekich, p. 40.

Chapter 6

1. Joel D. Wallach and Ma Lan, *The Agebeaters and Their Universal Currency for Immortality*, 2nd ed., Bonita, CA: Wellness Publications, 2011, p. vi.

2. Wallach and Lan, pp. x–xi.

3. Wallach and Lan, p. 3.

4. Wallach and Lan, pp. 17–18.

5. Wallach and Lan, pp. 20–24.

6. Wallach and Lan, pp. 36–51.

7. Wallach and Lan, pp. 51

8. David A. Kekich, *Life Extension Express*, Charleston, SC: Book Surge, 2010, p. 45.

9. Kekich, pp. 48–49.

10. Kekich, p. 73.

11. Kekich, pp. 54–55.

12. Kekich, pp. 55–58.

13. Kekich, pp. 58–60.

14. Kekich, p. 65.

15. Kekich, pp. 76–77.

16. Kekich, pp. 97–98.

17. Kekich, p. 99.

18. Kekich, p. 99.

19. Kekich, pp. 99–108.

20. James Lee, *The Methuselah Project*, North Charleston, SC: CreateSpace, 2014, pp. 6–28.

21. Lee, p. 28.

22. Lee, pp. 31–35.

23. Lee, pp. 37–41.

24. Lee, p. 47.

25. Lee, pp. 49–53.

26. Lee, pp. 53–55.

27. Lee, p. 61.

28. Lee, pp. 96–105.

29. "Overweight and Obesity Statistics," National Institute of Diabetes and Digestive Disorders, n.d. https://www.niddk.nih.gov/health-information/health-statistics/Pages/overweight-obesity-statistics.aspx.

30. Kay Uzoma, "Percentage of Americans Who Diet Each Year," LiveStrong.com, June 24, 2015. http://www.livestrong.com/article/308667-percentage-of-americans-who-diet-every-year/.

31. Jessica Migala, "33 Top Diet Plans That Are Actually Worth Trying," *Redbook*, December 12, 2016. http://www.redbookmag.com/body/health-fitness/features/g3134/top-diet-plans.

32. Ray Kurzweil and Terry Grossman, "Bridges to Life," in Gregory M. Fahy, Michael D. West, L. Stephen Coles, and Steven B. Harris, eds., *The Future of Aging*, Norco, CA: Springer Science & Business Media B.V., 2010, p. 337.

33. Kurzweil and Grossman, pp. 337–338.

34. Kurzweil and Grossman, pp. 339–341.

35. Kurzweil and Grossman, pp. 341–343.

36. Kurzweil and Grossman, p. 346.

37. "Dr. Walford's Interactive Diet Planner," Walford.com, n.d. http://www.walford.com/software.htm.

Chapter 7

1. James Lee, *The Methuselah Project*, North Charleston, SC: CreateSpace, June 6, 2014, p. 66.

2. Lee, pp. 66–67.

3. David A. Kekich, *Life Extension Express*, Charleston, SC: Book Surge, 2010, p. 81.

4. Kekichpp. 68, 81–84.

5. Kekich, p. 86.

6. Ray Kurzweil and Terry Grossman, "Bridges to Life," in F. M. Fahly, Michael D. West, L. Stephen Coles, and Steven B. Harris, eds., *The Future of Aging*, Norco, CA: Springer Science and Business Media B.V., 2010, pp. 360–361.

7. Kekich, pp. 87–88.

8. Kurzweil and Grossman, pp. 354–355.

9. Kekich, pp. 88–89.

10. Kekich, pp. 92–93.

11. Kurzweil and Grossman, p. 396.

12. National Institute on Aging, "Aging Hearts and Arteries," National Institutes of Health, National Institute on Aging, U.S. Department of Health and Human Services, 2005, n.d. https://www.nia.nih.gov/health/publication/aging-hearts-and-arteries/chapter-2-aging-heart.

13. Kekich, pp. 92–93.

14. Kurzweil and Grossman, p. 356.

15. Kurzweil and Grossman, pp. 367–398.

16. Kekich, p. 95.

17. Kurzweil and Grossman, pp. 349–350.

18. Kurzweil and Grossman, p. 351.

19. Kurzweil and Grossman, pp. 350–351.

20. Kekich, p. 135.

21. Kekich, pp. 125–138.

22. "Smoking and Tobacco Use," Centers for Disease Control and Prevention, December 1, 2016. https://www.cdc.gov/tobacco/data_statistics/fact_sheets/health_effects/tobacco_related_mortality/.

23. Kekich, p. 144.

24. Kekich, p. 147.

25. Kekich, p. 149.

26. Kekich, p. 154.

27. Lee, p. 78.

28. Kurzweil and Grossman, pp. 409–410.

29. Kekich, pp. 142–143.

30. Kurzweil and Grossman, pp. 408–409.

31. Kurzweil and Grossman, p. 410.

32. Kurzweil and Grossman, p. 412.

33. Kurzweil and Grossman, p. 412.

34. Kekich, pp. 178–179.

35. Norman Vincent Peale, *The Power of Positive Thinking*, reprint ed., New York: Touchstone, 2003.

36. Stephen Hawking. http://www.hawking.org.uk/about-stephen.html.

37. Nick Vukicic, *Life Without Limits*, Colorado Springs, CO: WaterBrook, 2012.

38. Gini Graham Scott, *Mind Power: Picture Your Way to Success*, West Nyack, NY: Parker Publishing Company, 1987.

39. Lee, p. 74.

40. Lee, p. 75.

41. Kekich, pp. 157–158.

42. Kekich, pp. 158–159.

43. Kekich, pp. 158, 159, 163.

44. Kurzweil and Grossman, pp. 174–175.

45. Kurzweil and Grossman, pp. 176–177.

46. Kurzweil and Grossman, pp. 182–183.

47. Kurzweil and Grossman, pp. 184–187.

Chapter 8

1. Ross Kenneth Urken, "Doctor Ready to Perform First Human Head Transplant," *Newsweek*, April 26, 2016. http://www.newsweek.com/2016/05/06/first-human-head-transplant-452240.html.

2. Urken.

3. Hannah Osborne, "Head Transplant: Servio Canavero, Announces Successful Repair of Spinal Cord," Newsweek, June 14, 2017. http://www.newsweek.com/head-transplant-sergio-canavero-repair-spinal-cord-rats-625689.

4. Ashley Welch, "Russian Man Volunteers for First Human Head Transplant," CBS News, August 29, 2016. http://www.cbsnews.com/news/russian-man-volunteers-for-first-human-head-transplant.

5. Welch.

6. Urken.

7. Bec Crew, "The Surgeon Behind the First Human Head Transplant Is Using VR to Prepare His Patients," ScienceAlert, November 21, 2016. http://www.sciencealert.com/the-surgeon-behind-the-first-human-head-transplant-will-use-vr-to-prepare-his-patients.

8. Urken.

9. Welch.

10. Richard Gray, Libby Plummer, and Abigail Beall, "Surgeon Behind World's First Human HEAD Transplant Says the Operation Could Take Place in the UK Next Year," *Daily Mail*, November 23, 2016. http://www.dailymail.co.uk/sciencetech/article-3965054/Surgeon-word-s-human-HEAD-transplant-says-operation-place-UK-year.html.

11. Gray, Plummer, and Beall.

12. Welch.

13. Blayne Hayes, "World's First Human Head Transplant to Take Place 2017," East Texas Matters, March 26, 2017. http://www.easttexasmatters.com/news/worlds-first-human-head-transplant-to-take-place-2017/681001972.

14. Lizbeth Aparicio, "World's First Human Head Transplant Set to Take Place in December 2017," *Affinity Magazine*, March 29, 2017. http://affinitymagazine.us/2017/03/29/worlds-first-human-head-transplant-set-to-take-place-in-december-2017/.

15. Hamilton Nolan, "Will We Ever Be Able to Transplant Human Brains?," Gawker, March 12, 2013. http://gawker.com/5990146/will-we-ever-be-able-to-transplant-human-brains.

16. Manuel Alfonseca, "Brain Transplant," *Popular Science* (blog), March 23, 2017. http://populscience.blogspot.com/2017/03/brain-transplant.html.

17. Nolan.

18. Nolan.

19. Nolan.

20. Nolan.

21. C. Ethier, E. R. Oby, M. J. Bauman, and L. E. Miller, "Restoration of Grasp Following Paralysis through Brain-Controlled Stimulation of Muscles," *Nature*, April 2012. http://www.nature.com/nature/journal/v485/n7398/full/nature 10987.html.

22. Sarah Knapton, "Humans Could Download Brains on to a Computer and Live Forever," *The Telegraph*, May 25, 2015. http://www.telegraph.co.uk/culture /hay-festival/11627328/Humans-could-download-brains-on-to-a-computer-and -live-forever.html.

23. Knapton.

24. Chirp, "Mind Transfer to a Computer Could Be Possible by 2050. Immortality May Be Within Reach," *Chirp News*, April 5, 2016. http://chirpnews.com /2016/04/05/mind-transfer-to-computer.

25. Chirp.

26. Chirp.

27. Jordan Inafuku. https://cs.stanford.edu/people/eroberts/cs181/projects /2010-11/DownloadingConsciousness/tandr.html, "Downloading Consciousness," Stanford Computer Science. https://cs.stanford.edu/people/eroberts/cs181/projects /2010-11/DownloadingConsciousness/tandr.html.

28. Inafuku et al.

29. Chirp.

30. "The Immortalist: Unloading the Mind to a Computer," BBC, March 14, 2016. http://www.bbc.com/news/magazine-35786771.

31. "The Immortalist."

32. "The Immortalist."

33. "The Immortalist."

34. Inafuku et al.

35. Inafuku et al.

36. "Brain Transplant."

37. *Ghost in the Shell*, March 31, 2107 release date. http://www.imdb.com/title /tt1219827/?ref_=fn_al_tt_1.

Chapter 9

1. Larry Schwartz, "4 Ways the One Percent Is Trying to Buy Their Immortality," Alternet, June 12, 2015. http://www.alternet.org/print/personal-health/4 -ways-one-percent-trying-buy-their-immortality.

2. Ariana E. Cha, "Tech Titans' Latest Project: Defy Death," *Washington Post*, April 4, 2016. http://www.washingtonpost.com/sf/national/2015/04/04/tech -titans-latest-project-defy-death/?utm_term=.78be71e394db.

3. Cha.

4. Schwartz.

5. Schwartz.

6. Reuters, "Genetic Tinkering Is Found to Extend Roundworms Lives," March 8, 2001. http://www.nytimes.com/2001/03/08/us/genetic-tinkering-is-found-to-extend-roundworms-lives.html.

7. Betsy Isaacson, "Silicon Valley Is Trying to Make Humans Immortal—And Finding Some Success," *Newsweek*, March 5, 2015. http://www.newsweek.com/2015/03/13/silicon-valley-trying-make-humans-immortal-and-finding-some-success-311402.html.

8. Isaacson.

9. Madhumita Murgia, "Will Technology Help Us Live Forever?," *The Telegraph*, January 21, 2016. http://www.telegraph.co.uk/technology/2016/01/25/will-technology-help-us-live-forever.

10. Murgia.

11. Zoe Corbyn, "Live for Ever: Scientists Say They'll Soon Extend Life 'Well Beyond 120,'" *The Guardian*, January 11, 2015. https://www.theguardian.com/science/2015/jan/11/-sp-live-forever-extend-life-calico-google-longevity.

12. Isaacson.

13. Corbyn.

14. Katrina Brooker, "Google Ventures and the Search for Immortality," Bloomberg, March 8, 2015. https://www.bloomberg.com/news/articles/2015-03-09/google-ventures-bill-maris-investing-in-idea-of-living-to-500.

15. Brooker.

16. Corbyn.

17. Murgia.

18. Ronald Klatz, "New Horizons for the Clinical Specialty of Anti-aging Medicine: The Future with Biomedical Techologies," *Annals of the New York Academy of Sciences* 1057 (December 2005), pp. 536–544. http://onlinelibrary.wiley.com/wol1/doi/10.1196/annals.1356.041/abstract.

19. Klatz.

20. Klatz.

21. Klatz.

22. Isaacson.

23. Ian Sample, "Can We Reverse the Ageing Process by Putting Young Blood into Older People?," *The Guardian*, August 4, 2015. https://www.theguardian.com/science/2015/aug/04/can-we-reverse-ageing-process-young-blood-older-people.

24. Sample.

25. Isaacson.

26. Isaacson.

27. Sample.

28. Sample.

29. Sarah C. P. Williams, "Old-Age Protein May Cause Memory Loss," Science, July 6, 2015. http://www.sciencemag.org/news/2015/07/old-age-protein-may-cause-memory-loss.

30. Sample.

31. Isaacson.

32. Natalie Kalin, "Immortality May Be More Than Mere Fiction," HuffPost, February 9, 2016. http://www.huffingtonpost.com/natalie-kalin/immortality-may-be-more-t_b_9178214.html.

33. Gary Vey, "Can Science Make Us Immortal?," Viewzone, n.d. http://www.viewzone.com/aging.html.

34. Kalin.

35. A. G. Bodnar, M. Ouellette, M. Frolkis, S. E. Holt, C. P. Chiu, G. B. Morin, C. B. Harley, J. W. Shay, S. Lichtsteiner, and W. E. Wright, "Extension of Life-Span by Introduction of Telomerase into Normal Human Cells," *Science*, New Series 279 (January 16, 1998), pp. 349–352.

36. Bodnar et.al.

37. Isaacson.

38. Isaacson.

39. Dan Krotz, "Scientists Develop New Way to Study How Human Cells Become Immortal, a Crucial Precursor to Cancer," News Center, November 6, 2014. http://newscenter.lbl.gov/2014/11/06/cancer/.

40. Kalin.

41. Kalin.

42. Adam Leith Gollner, "Live Forever! Can Science Deliver Immortality?," Salon.com, August 18, 2013. http://www.salon.com/2013/08/18/live_forever_can_science_deliver_immortality.

43. Douglas Perry, "Want to Live Forever? New Research Suggests Science Could Extend the Typical Human Lifespan Indefinitely," OregonLive, January 12, 2015. http://www.oregonlive.com/living/index.ssf/2015/01/want_to_live_forever_new_resea.html.

44. Gollner.

45. Gollner.

46. Gollner.

47. Gollner.

48. Gollner.

49. Isaacson.

50. Isaacson.

51. Isaacson.

References

Alfonseca, Manuel. "Brain Transplant." *Popular Science* (blog), March 23, 2017. http://populscience.blogspot.com/2017/03/brain-transplant.html.

Aparicio, Lizbeth. "World's First Human Head Transplant Set to Take Place in December 2017." *Affinity Magazine*, March 29, 2017. http://affinitymaga zine.us/2017/03/29/worlds-first-human-head-transplant-set-to-take-place -in-december-2017/.

Blair, S. N., H. W. Kohl 3rd, R. S. Paffenbarger Jr, D. G. Clark, K. H. Cooper, and L.W. Gibbons. "Physical Fitness and All-Cause Mortality: A Prospective Study of Healthy Men and Women." *Journal of the American Medical Association* 262 (1989), pp. 2395–2401.

Bodnar, Andrea G., Michel Ouellette, Maria Frolkis, Shawn E. Holt, Choy-Pik Chiu, Gregg B. Morin, Calvin B. Harley, Jerry W. Shay, Serge Lichtsteiner, and Woodring E. Wright. "Extension of Life-Span by Introduction of Telomerase into Normal Human Cells." *Science, New Series* 279, no. 5349 (October 12, 2016), pp. 349–352.

Brooker, Katrina. "Google Ventures and the Search for Immortality." Kurzweilai. net, March 9, 2015. http://www.kurzweilai.net/bloomberg-google-ventures -and-the-search-for-immortality.

Burzynski, Stanislaw. "Practical Application of Gene Silencing Theory of Aging: Life Extension in Animal Testing and Human Clinical Trials." *Anti-Ageing Medical Therapeutics* 11 (January 2009), pp. 1–8.

Carter, Chris. *Science and the Afterlife Experience.* Rochester, VT: Inner Traditions, 2012.

Cha, Ariana E. "Tech Titans' Latest Project: Defy Death." *Washington Post*, April 4, 2015. http://www.washingtonpost.com/sf/national/2015/04/04/tech-titans -latest-project-defy-death/?utm_term=.684125a9b22b.

Chirp. "Mind Transfer to a Computer Could Be Possible by 2050. Immortality May Be Within Reach." *Chirp News*, April 5, 2016. http://chirpnews.com /2016/04/05/mind-transfer-to-computer.

Collman, Ashley. "Revealed: Red Sox Legend Ted Williams Was Cryogenically Frozen AGAINST HIS WILL by His Son in the Hope He Could Be Brought

Back to Life Someday." *Daily Mail*, December 1, 2013. http://www.daily mail.co.uk/news/article-2516442/Red-Sox-legend-Ted-Williams-cryogen ically-frozen-AGAINST-HIS-WILL.html.

Corbyn, Zoe. "Live for Ever: Scientists Say They'll Soon Extend Life 'Well Beyond 120.'" *The Guardian*, January 11, 2015. https://www.theguardian.com/sci ence/2015/jan/11/-sp-live-forever-extend-life-calico-google-longevity.

Crew, Bec. "The Surgeon Behind the First Human Head Transplant Is Using VR to Prepare His Patients." ScienceAlert, November 21, 2016. http://www .sciencealert.com/the-surgeon-behind-the-first-human-head-transplant-will -use-vr-to-prepare-his-patients.

"Dr. Walford's Interactive Diet Planner." Walford.com, n.d. http://www.walford.com /software.htm.

Ethier, C., E. R. Oby, M. J. Bauman, and L. E. Miller. "Restoration of Grasp Fol-lowing Paralysis through Brain-Controlled Stimulation of Muscles." *Nature*, April 2012. http://www.nature.com/nature/journal/v485/n7398 /full/nature10987.html?foxtrotcallback=true.

Ettinger, Robert C. W. *The Prospect of Immortality*. Painesville, OH: Ria University Press, 2005.

"February 10, 2012: Lawsuits against Larry Johnson End; Johnson Issues State-ment." Alcor, n.d. http://www.alcor.org/blog/lawsuits-against-larry-johnson -end-johnson-issues-statement/.

"Generation Cryo: Fighting Death in the Frozen Unknown." Gizmodo.com, October 5, 2016. http://gizmodo.com/generation-cryo-fighting-death-in -the-frozen-unknown-1786446378.

Gollner, Adam Leith. "Live Forever! Can Science Deliver Immortality?" Salon. com, August 18, 2013. http://www.salon.com/2013/08/18/live_forever_can _science_deliver_immortality.

Gray, Richard, Libby Plummer, and Abigail Beall. "Surgeon Behind World's First Human HEAD Transplant Says the Operation Could Take Place in the UK Next Year." *Daily Mail*, November 23, 2016. http://www.dailymail.co .uk/sciencetech/article-3965054/Surgeon-word-s-human-HEAD-transplant -says-operation-place-UK-year.html.

Grunberg, Alexandra. "From Cyborgs to Nanobots: 5 Ways Scientists Hope to Achieve Immortality for Humanity." Outerplaces.com, July 20, 2015. https://www.outerplaces.com/science/item/9395-from-cyborgs-to-nanobots -5-ways-scientists-hope-to-achieve-immortality.

Hart, Hornell. *The Enigma of Survival*. London: Rider, 1959.

Hayes, Blayne. "World's First Human Head Transplant to Take Place 2017." East Texas Matters, March 26, 2017. http://www.easttexasmatters.com/news /worlds-first-human-head-transplant-to-take-place-2017/681001972.

Hellwig, Heinrik. "The Immortality Project, University of California, Riverside John Martin Fischer, Project Leader." *John Templeton Foundation* (August 21, 2016), pp. 1–43.

"The Immortalist: Unloading the Mind to a Computer." BBC, March 14, 2016. http://www.bbc.com/news/magazine-35786771.

Inafuku, Jordan, Katie Lampert, Brad Lawson, Shaun Stehly, and Alex Vaccaro. "Downloading Consciousness." Stanford Computer Science. https://cs .stanford.edu/people/eroberts/cs181/projects/2010-11/Downloading Consciousness/tandr.html.

Isaacson, Betsy. "Silicon Valley Is Trying to Make Humans Immortal—And Finding Some Success." Newsweek.com, March 9, 2015. http://www .newsweek.com/2015/03/13/silicon-valley-trying-make-humans-immortal -and-finding-some-success-311402.html.

Johnson, Larry. *Frozen: My Journey into the World of Cryonics, Deception, and Death.* New York: Vanguard Press, 2009.

"Juan Ponce de León." History Channel, n.d. http://www.history.com/topics/explo ration/juan-ponce-de-leon.

Kalin, Natalie. "Immortality May Be More Than Mere Fiction." HuffPost, February 9, 2016. http://www.huffingtonpost.com/natalie-kalin/immortality-may -be-more-t_b_9178214.html.

Kekich, David A. *Life Extension Express.* Newport Beach, CA: Booksurge Publishing, 2009.

Klatz, Ronald. "New Horizons for the Clinical Specialty of Anti-aging Medicine: The Future with Biomedical Technologies." *Annals of the New York Academy of Sciences* 1057 (2005), pp. 536–544. https://www.ncbi.nlm.nih .gov/pubmed/16399918.

Knapton, Sarah. "Humans Could Download Brains on to a Computer and Live Forever." *The Telegraph*, May 25, 2015. http://www.telegraph.co.uk/culture /hay-festival/11627328/Humans-could-download-brains-on-to-a-computer -and-live-forever.html.

Krotz, Dan. "Scientists Develop New Way to Study How Human Cells Become Immortal, a Crucial Precursor to Cancer." News Center, November 6, 2014. http://newscenter.lbl.gov/2014/11/06/cancer/.

Kurzweil, Ray, and Terry Grossman. "Bridges to Life." In G. M. Fahely, M. D. West, L.S. Coles, and S. B. Harris (eds.), *The Future of Aging.* New York: Springer Science & Business Media B.V., 2010, pp. 3–22.

Lane, Mark A., Julie Mattison, Donald K. Igram, and George S. Roth. "Caloric Restriction and Aging in Primates: Relevance to Humans and Possible CR Mimetics." *Microscopy Research and Technique* 59(2002), pp. 335–338.

Lee, James. *The Methuselah Project.* Berkeley, CA: CPSIA, 2014.

Lee, Joel. "Want to Live Forever? 6 Technologies That Could Stop Aging." MakeUseOf, August 12, 2015. http://www.makeuseof.com/tag/want-live -forever-6-technologies-eliminate-aging/.

McCormack, David. "We Did It Out of Love: Baseball Legend Ted Williams' Daughter Finally Speaks Out about Why She and Her Brother Spent $100,000 to Have Their Father's Body Cryogenically Frozen." *Daily Mail*, May 19, 2014.

http://www.dailymail.co.uk/news/article-2632809/We-did-love-Baseball
-legend-Ted-Williams-daughter-finally-speaks-brother-spent-100-000
-fathers-body-cryogenically-frozen.html.

Migala, Jessica. "33 Top Diet Plans That Are Actually Worth Trying." *Redbook*,
December 12, 2016. http://www.redbookmag.com/body/health-fitness
/features/g3134/top-diet-plans.

Murgia, Madhumita. "Will Technology Help Us Live Forever?" *The Telegraph*,
January 21, 2016. http://www.telegraph.co.uk/technology/2016/01/25/will
-technology-help-us-live-forever/.

National Institute on Aging. "Aging Hearts and Arteries." National Institute of
Aging, n.d. https://www.nia.nih.gov/health/publication/aging-hearts-and
-arteries/chapter-2-aging-heart.

Nelson, Bob. *Freezing People Is (Not) Easy: My Adventures in Cryonics*. Guilford,
CT: Lyons Press, 2014.

Nolan, Hamilton. "Will We Ever Be Able to Transplant Human Brains?" Gawker,
March 12, 2013. http://gawker.com/5990146/will-we-ever-be-able-to-trans
plant-human-brains.

"Overweight and Obesity Statistics." National Institute of Diabetes and Digestive
Disorders, n.d. https://www.niddk.nih.gov/health-information/health-sta
tistics/Pages/overweight-obesity-statistics.aspx.

Peale, Norman Vincent. *The Power of Positive Thinking*, reprint ed. New York:
Touchstone, 2003.

Perry, Douglas. "Want to Live Forever? New Research Suggests Science Could
Extend the Typical Human Lifespan Indefinitely." OregonLive, January 12,
2015. http://www.oregonlive.com/living/index.ssf/2015/01/want_to_live
_forever_new_resea.html.

Perry, Michael. "Our Finest Hours: Notes on the Dora Kent Crisis." Cryonics,
December 9, 1987–January 12, 1988. http://www.alcor.org/Library/html
/DoraKentCase.html.

"Problems Associated with Cryonics (and Some Possible Solutions)." Alcor, n.d.
http://www.alcor.org/problems.html.

"Response to Larry Johnson Allegations." Alcor, n.d. http://www.alcor.org/press
/response.html.

Sample, Ian. "Can We Reverse the Ageing Process by Putting Young Blood into
Older People?" *The Guardian*, August 4, 2015. https://www.theguardian
.com/science/2015/aug/04/can-we-reverse-ageing-process-young-blood
-older-people.

Sandomir, Richard. "Fight over Williams' Frozen Body May End Soon." *New York
Times*, September 26, 2002. http://www.nytimes.com/2002/09/26/sports
/fight-over-williams-s-frozen-body-may-end-soon.html.

Schwartz, Larry. "4 Ways the One Percent Is Trying to Buy Their Immortality."
Alternet, June 12, 2015. http://www.alternet.org/personal-health/4-ways
-one-percent-trying-buy-their-immortality.

Scott, Gini Graham. *Mind Power: Picture Your Way to Success*. Englewood Cliffs,
NJ: Prentice Hall, 1987.

Smith, Mat. "Japanese Latest Humanoid Robot Makes Its Own Moves." Engadget, July 30, 2016. https://www.engadget.com/2016/07/30/japan-humanoid
-alter-robot.

"Smoking and Tobacco Use." Centers for Disease Control and Prevention, December 1, 2016. https://www.cdc.gov/tobacco/data_statistics/fact_sheets/health
_effects/tobacco_related_mortality/.

Stephen Hawking. http://www.hawking.org.uk/about-stephen.html.

"Top 10 Ways You Could Live Forever." Listverse, January 1, 2013. http://listverse
.com/2013/01/01/top-10-ways-you-could-live-forever.

"Turning Off 'Aging Genes.'" ScienceDaily, January 2, 2014. https://www.science
daily.com/releases/2014/01/140102123403.htm.

Urban, Tim. "Why Cryonics Makes Sense." WaitButWhy, March 24, 2016. http://
www.waitbutwhy.com/2016/03/cryonics.html.

Urken, Ross Kenneth. "Doctor Ready to Perform First Human Head Transplant."
Newsweek, April 26, 2016. http://www.newsweek.com/2016/05/06/first
-human-head-transplant-452240.html.

Uzoma, Kay. "Percentage of Americans Who Diet Each Year." LiveStrong.
com, June 24, 2015. http://www.livestrong.com/article/308667-percentage
-of-americans-who-diet-every-year.

Vasquez, Laurie-Anne. "How Science Is Making Immortality a Reality." *Fiat
Physica* (blog), April 6, 2015. https://www.fiatphysica.com/blog/making
-history/science-of-immortality-aging-wrinkles.

Vey, Gary. "Can Science Make Us Immortal?" Viewzone, n.d. http://www.viewzone
.com/aging.html.

Vukicic, Nick. *Life Without Limits*. Colorado Springs, CO: WaterBrook, 2012.

Wallach, Joel D., and Ma Lan. *The Agebeaters and Their Universal Currency for
Immortality*. Chula Vista, CA: Wellness Publications, 2008.

Wallach, Joel D., and Ma Lan. *Immortality*, 2nd ed. Bonita, CA: Wellness Publications, 2011.

Welch, Ashley. "Russian Man Volunteers for First Human Head Transplant."
CBS News, August 29, 2016. http://www.cbsnews.com/news/russian-man
-volunteers-for-first-human-head-transplant.

"Welcome to the CR Society International!" CR Society International, n.d. http://
www.crsociety.org.

Yusuf, Salim, Steven Hawken, Stephanie Ôunpuu, Tony Dans, Alvaro Avezum,
Fernando Lanas, Matthew McQueen, Andrzej Budjai, Prem Pais, John
Varigos, and Liu Lisheng. "Effect of Potentially Modifiable Risk Factors
Associated with Myocardial Infarction in 52 Countries (the INTER-
HEART Study): Case-Control Study." *The Lancet*, September 11, 2004.
http://www.thelancet.com/journals/lancet/article/PIIS0140-6736(04)17018
-9/abstract.

Index

About the Author

GINI GRAHAM SCOTT, PHD, JD, is a nationally known writer, a consultant, speaker, and seminar leader, specializing in social trends, popular culture, business and work relationships, and professional and personal development. She has authored dozens of books on diverse subjects. She has assisted dozens of clients on memoirs, self-help, and popular business books as well as film scripts. She is a *Huffington Post* regular columnist, commenting on social trends, new technology, business, and everyday life at http://www.huffingtonpost.com/gini-graham-scott.

She has written or coauthored five recent books, including *How the Rich Kill; Lies and Liars; Scammed; American Justice?* with Paul Brakke; *Credit Card Fraud* with Jennifer Grondahl Lee; and *At Death's Door* with Sebastian Sepulveda. *At Death's Door* is expected to be produced as a TV pilot for a new series to be rolled out after the book's publication.

She has received national media exposure for her own books, including appearances on *Good Morning America, Oprah, Montel Williams,* and CNN. She has been the producer and host of a talk show series, *Changemakers*, featuring interviews on social trends.

Scott is active in a number of community and business groups, including the Lafayette, Pleasant Hill, and Danville Chambers of Commerce. She is a graduate of the Leadership in Contra Costa County program and is a member of a BNI group in the East Bay, B2B groups in Danville and Walnut Creek, and the Lafayette Women's Connection. She is the organizer of six Meetup groups in the film and publishing industries with over 6,000 members in Los Angeles and the San Francisco Bay Area. She also does workshops and seminars on the topics of her books and on self-publishing.

She received her PhD from the University of California, Berkeley, and her JD from the University of San Francisco Law School. She has received several MA degrees from Cal State, East Bay, most recently an MA in Communications. She will be getting an additional MA in history starting in September 2017.